Introduction to Mathematical Portfolio Theory

In this concise yet comprehensive guide to the mathematics of modern portfolio theory, the authors discuss mean–variance analysis, factor models, utility theory, stochastic dominance, very long term investing, the capital asset pricing model, risk measures including VAR, coherence, market efficiency, rationality and the modelling of actuarial liabilities. Each topic is clearly explained with assumptions, mathematics, limitations, problems and solutions presented in turn.

Joshi's trademark style of clarity and practicality is here brought to classical financial mathematics. The book is suitable for mathematically trained students in actuarial studies, business and economics as well as mathematics and finance, and it can be used both for self-study and as a course text. The authors' experience as both academics and practitioners brings clarity and relevance to the book, whilst ensuring that the limitations of models are highlighted.

MARK S. JOSHI is a researcher and consultant in mathematical finance, and a Professor at the University of Melbourne. His research focuses on derivatives pricing and interest rate derivatives in particular. He is the author of numerous research articles on quantitative finance and four books.

JANE M. PATERSON obtained a PhD in pure mathematics from the University of Melbourne. She furthered her academic experience with a postdoctoral fellowship at the Mathematical Sciences Research Institute, Berkeley and a research fellowship at the University of Cambridge. More recently she has worked in both the UK and Australia as a director in a variety of specialist and generalist banking roles, including structured finance and economic capital, with organisations including National Australia Bank and ANZ.

INTERNATIONAL SERIES ON ACTUARIAL SCIENCE

The *International Series on Actuarial Science*, published by Cambridge University Press in conjunction with the Institute and Faculty of Actuaries, contains textbooks for students taking courses in or related to actuarial science, as well as more advanced works designed for continuing professional development or for describing and synthesising research. The series is a vehicle for publishing books that reflect changes and developments in the curriculum, that encourage the introduction of courses on actuarial science in universities, and that show how actuarial science can be used in all areas where there is long-term financial risk.

A complete list of books in the series can be found at www.cambridge.org/statistics. Recent titles include the following:

Solutions Manual for Actuarial Mathematics for Life Contingent Risks (2nd Edition)
David C.M. Dickson, Mary R. Hardy & Howard R. Waters

Actuarial Mathematics for Life Contingent Risks (2nd Edition)
David C.M. Dickson, Mary R. Hardy & Howard R. Waters

Risk Modelling in General Insurance
Roger J. Gray and Susan M. Pitts

Financial Enterprise Risk Management
Paul Sweeting

Regression Modeling with Actuarial and Financial Applications
Edward W. Frees

Nonlife Actuarial Models
Yiu-Kuen Tse

Generalized Linear Models for Insurance Data
Piet De Jong & Gillian Z. Heller

Market-Valuation Methods in Life and Pension Insurance
Thomas Møller & Mogens Steffensen

INTRODUCTION TO MATHEMATICAL PORTFOLIO THEORY

MARK S. JOSHI
JANE M. PATERSON

CAMBRIDGE UNIVERSITY PRESS
Cambridge, New York, Melbourne, Madrid, Cape Town,
Singapore, São Paulo, Delhi, Mexico City

Cambridge University Press
The Edinburgh Building, Cambridge CB2 8RU, UK

Published in the United States of America by Cambridge University Press, New York

www.cambridge.org
Information on this title: www.cambridge.org/9781107042315

First published 2013

Printed and bound in the United Kingdom by Bell and Bain Ltd

A catalogue record for this publication is available from the British Library

ISBN 978-1-107-04231-5 Hardback

Contents

Preface

This book grew out of a lecture course taught at the University of Melbourne over a series of years. The audience was third-year actuarial students who partially gained an exemption from the Faculty and Institute of Actuaries CT8 module if they did well. The nature of the audience and the exemption placed certain constraints on the syllabus and delivery that made it hard to find a suitable textbook. The graduate level texts simply being too hard, whilst the undergraduate and MBA books did not cover the mathematics in sufficient depth. In particular, the students were fairly mathematical but more oriented towards computations than proof. In addition, the choice of topics had to be tuned to the actuarial syllabus and that is reflected in this book.

In terms of mathematical level, we strive to achieve a mid-level where mathematics is not shied away from nor hived off to appendices, but also not so hard as to deter the undergraduate reader. Also, this being a book on mathematical portfolio theory, the mathematics takes centre stage for most of the book: our objective is to study the mathematics of portfolio theory without losing sight of the finance. As both authors have been both practitioners and academics, a theme throughout is that a model is a model and not reality, and we aim to highlight our assumptions and their consequences. We provide a lot of problems with solutions in the belief that this is ultimately how the material is best learnt, and as a consequence of the fact that students always want more problems and more solutions.

We first look at the definitions of risk and return. We then explore Markowitz's portfolio theory. We start with the two-asset case, then add a riskless asset, and finally treat the general case. We derive a couple of different ways to find efficient portfolios in that case. We then move on to seeing how simplifying correlation structures can help to reduce the amount of data needed to estimate the model parameters.

We then make a long excursion into utility theory, looking at both its pros and cons. We also look at the offshoots of stochastic dominance and geometric mean maximisation. An important issue in any insurance company or bank is risk control, and we therefore look at risk measures including VAR and conditional expected shortfall. We also examine the coherence axioms.

We then move onto a critical look at the capital asset pricing model and the arbitrage pricing theory. We also discuss market efficiency and rationality. Here we adopt a more discursive viewpoint. We finish by looking at long-term models of stock prices using Brownian motion and the Wilkie model.

Much of this book has been shaped by interactions with our students and from their explicit feedback, and we thank them all for their input. We particularly thank Timothy Hillman for his detailed comments on the manuscript.

For the reader whose appetite has been whetted and wishes to study the material in greater depth, we mention a few books which we have found helpful.

- Elton *et al.*, [6], is very discursive and contains very detailed references but is much less mathematical than this book.
- Pennachi, [15], is a nice text aimed at PhD students.
- Cochrane, [3], has both good discussion and good mathematics and will reward the reader who perseveres.
- Markowitz's original book, *Portfolio Selection,* [11], is still a good read.

Mark Joshi
Jane Paterson
Melbourne 2012

1
Definitions of risk and return

1.1 Introduction

Mathematics can be applied to the practice of finance in a number of ways. These include trying to use mathematics to predict asset price movements (*statistical arbitrage*), measuring and controlling risk in trading books (*risk management,*) pricing options and other contingent claims by assessing hedging strategies (*derivatives pricing,*) and the use of mathematics to maximise the risk–return trade-off when investing in the markets: *portfolio theory*. It is portfolio theory which we will address in this volume. This subject is sometimes called *modern portfolio theory* or *MPT*.

At first glance the objective of maximising the risk–return trade-off when investing in the markets appears straightforward and intuitive; that is, any rational investor will want to maximise the anticipated return on his or her investment whilst minimising the risk of unexpected loss. However, in order to apply the rigour of mathematics to this activity, we need first to carefully define these terms. What is risk? What is return? How do we decide the trade-off between them? There are multiple ways of doing this, and we will examine the more widely-used ones.

Throughout we will make two fundamental assumptions. The first is that individual assets are correctly priced. This means that "stock picking" is pointless and, accordingly, our efforts will focus on how to compare portfolios with each other.

Second, to ensure consistency, we will generally work across a fixed timeframe, for example, one year. We should think of ourselves as a funds manager whose performance is assessed on a yearly basis. The funds manager will be given a statement by his/her client or the board stating the required risk–return trade-off and then it becomes his or her job to achieve it.

1.2 Measuring return

The return on an asset is the percentage change in its value over a given time period. A negative return is possible (and the financial markets upheavals of 2008 and beyond have seen many negative returns). Notably, change can occur in multiple fashions. For example, in the case of a stock, its market price can vary both up and down due to company performance and general market conditions. Second, the stock may pay dividends which will always be considered as part of the return. Dividends may be paid either in cash, or as *scrip* dividends in the form of additional shares. For a scrip dividend the number of shares held by the investor increases but no cash changes hands; value increases and the contribution to return is positive. By way of contrast, certain exotic financial instruments as well as real-world assets require the holder to pay money in order to retain his or her rights: this results in negative cash-flows which must also be considered as part of the return.

Definition 1.1 The *return* on a portfolio is the percentage change in its value taking into account all cash in-flows and out-flows.

In financial mathematics, we are generally interested in the future rather than the past, so the return will normally be uncertain. It is therefore *expected* return that is important rather than *actual* return. We therefore assume that the return, R, follows some probability distribution with density f. And the *expected return* is

$$\mathbb{E}(R) = \overline{R} = \int R f(R) dR.$$

In simple examples, R will be a discrete random variable, and f is then a sum of point densities (delta functions) and R follows some probability distribution taking values R_i with probability p_i.

The *expected return* is then

$$\mathbb{E}(R) = \overline{R} = \sum_{i=1}^{n} p_i R_i.$$

For example, if R has probabilities of taking values as follows

p_i	R_i
$\frac{1}{3}$	5%
$\frac{1}{6}$	6%
$\frac{1}{2}$	7%

,

then the expected percentage return is

$$\frac{1}{3}5+\frac{1}{6}6+\frac{1}{2}7.$$

The expectation operator is linear; that is

$$\mathbb{E}(aX+bY)=a\mathbb{E}(X)+b\mathbb{E}(Y),$$

and so for a portfolio P consisting of assets A_i, with return R_i, in proportions X_i, we have

$$\mathbb{E}(R_P)=\overline{R}_P=\sum_{i=1}^{n}X_i\overline{R}_i.$$

We can write this as

$$\overline{R}_P=\langle X,\overline{R}\rangle, \tag{1.1}$$

with $X=(X_1,\ldots,X_n)$ and $\overline{R}=(\overline{R}_1,\ldots,\overline{R}_n)$.

Here $\langle x,y\rangle$ denotes the inner product of two vectors: $\sum_{i=1}^{n} x_i y_i$.

1.3 Portfolio constraints

Generally, when given the task of building an investment portfolio, we will be entrusted with a fixed sum of money to divide between a total of n possible assets, $A_1,\ldots A_n$. We will solely be interested in what fraction of our money to put into each asset and hence we can take the sum of money to be 1. If we put X_i into each asset our constraint becomes

$$\sum_{i=1}^{n}X_i=1. \tag{1.2}$$

What further constraints do we put on X_i? Often, there is a requirement that all holdings be non-negative. In other words, we are not allowed to *short-sell* assets. Effective short-selling is possible in the markets under certain constraints, although in the wake of the 2008 global financial crisis, various regulators intervened to ban the short-selling of financial stocks in attempts to control speculation and spiraling confidence loss. The long-term efficacy of this strategy has been a topic of some debate and in any case, derivatives can generally be used to obtain the same effects. For a portfolio prohibiting short-sales, we will have the extra constraint

$$X_i\geq 0.$$

It is also not unusual for funds to place restrictions on the fraction of wealth that can be put into a single asset so often there will be a single asset constraint of the form

$$X_i \leq \varepsilon_i,$$

for some given ε_i.

Other commonly-encountered portfolio constraints include restrictions on the fraction of wealth to be invested in particular geographies, industries, credit ratings or asset classes. These may be set to reflect preferences of the board or client as the case may be and will vary according to the investment strategy of a particular fund or portfolio.

Note that if our sole objective is to maximise expected return, the portfolio selection problem is easy to solve. We simply put as much money as possible into the highest returning asset. In other words, we find the asset, A_j, that maximises $\overline{R}_j$ and invest all the money in that. If there is a constraint on how much money can be placed in each asset, then we put as much as we can in A_j and then as much of what is left as we can in the next best and so on.

The reason that there is some work (and mathematics) to this subject is that generally there is a requirement to control risk to some stated level of "risk appetite" as well as to maximise returns. To control risk we first need to define it and to guide our thinking, we start by looking at some very simple examples.

Example 1.2 Suppose we have to choose between two assets. Asset A pays \$ 1,000,000 with 25% probability and pays 0 with 75% probability. Asset B however pays \$ 250,000 with 100% probability.

Which would an investor prefer?

Both assets have the same mean of \$ 250,000. However, B guarantees the investor will receive the mean whereas A involves a great deal of risk.

Generally then, B would be preferred as it involves no risk. ◇

Example 1.3 Suppose that now we have to choose between two assets as follows. Asset A pays \$ 1,000,000 with 25% probability and pays 0 with 75% probability. Asset B pays \$ 260,000 with 100% probability.

Which would an investor prefer?

Asset B has higher mean and lower risk. You would have to be very risk-loving to prefer A. Note, however, that if you play roulette or a lottery, then A is the sort of investment you are making. Of course, owning a casino or running a lottery is a different matter and is highly recommended.

Incidentally, the mathematician's way of playing the lottery is to pick some numbers and then not buy a ticket. When the numbers do not come up, he has won a dollar. ◇

Whilst the above two examples are straightforward, the next raises some more interesting points.

Example 1.4 Asset A pays \$ 1,000,000 with 25% probability and pays 0 with 75% probability. Asset B however pays \$ 240,000 with 100% probability.

Which would an investor prefer?

Asset B has lower mean and lower risk. Most investors would prefer B on the grounds that the extra risk is not worth the extra \$10,000 to be gained on average. However, some might go for A since if one had the opportunity to do many such independent investments then the risk would average out (through diversification) and A would be preferable. This is in essence what banks do when they make a very large number of small loans to different customers and is often referred to as taking a "portfolio view." ◇

Example 1.5 Now suppose we have two assets A and B. A coin is tossed and A pays 1 on heads and zero otherwise. Asset B pays 1 on tails and zero otherwise. The two assets are based on the same coin toss. How much are A and B worth?

The mean pay-off for each asset is 0.5, yet we would expect the value to be lower because of risk aversion. We would also expect the two assets to trade at the same price. However, if we consider the portfolio of A and B together then it will always be worth 1 (the assets are complementary). We therefore conclude that the individual assets are worth 0.5 despite risk aversion.

This example illustrates the fact that a risk premium is generally not available for risk that is diversifiable or hedgeable. Whilst we will generally be unable to remove all risk, we will be able to remove some via portfolio diversification. ◇

1.4 Defining risk with variance

There are many ways to define and control risk. The first and simplest way is to use variance. The variance of a random variable is defined via

$$\mathrm{Var}(R) = \mathbb{E}((R - \overline{R})^2) = \mathbb{E}(R^2) - \mathbb{E}(R)^2.$$

The standard deviation is a related measure of risk. It is defined by

$$\sigma_R = (\mathrm{Var}\, R)^{\frac{1}{2}}.$$

It therefore contains the same information as the variance.

Standard deviation is harder to work with computationally because of the square root, but has the virtue that it has the same scale as the expectation. That is, we have

$$\begin{aligned}\mathbb{E}(\lambda R) =& \lambda \mathbb{E}(R), \\ \operatorname{Var}(\lambda R) =& \lambda^2 \operatorname{Var}(R), \\ \sigma_{\lambda R} =& |\lambda| \sigma_R.\end{aligned}$$

Note the important modulus sign, $||$, in the final equation: standard deviation is never negative.

We will be interested in the variance of portfolio returns given the variances of individual asset's returns. If we have assets with returns $R_1, \ldots, R_n$, held in amounts $X_1, \ldots, X_n$ then we can compute the variance of the portfolio.

We proceed by direct computation. We seek the value of

$$\operatorname{Var}(R_P) = \operatorname{Var}\left(\sum_{i=1}^{n} X_i R_i\right).$$

We compute

$$\begin{aligned}\operatorname{Var}\left(\sum_{i=1}^{n} X_i R_i\right) =& \operatorname{Var}\left(\sum_{i=1}^{n} X_i (R_i - \mathbb{E}(R_i))\right), \\ =& \mathbb{E}\left(\left(\sum_{j=1}^{n} X_j (R_j - E(R_j))\right)^2\right), \\ =& \mathbb{E}\left(\sum_{i=1}^{n} X_i (R_i - \mathbb{E}(R_i)). \sum_{j=1}^{n} X_j (R_j - \mathbb{E}(R_j))\right), \\ =& \sum_{i,j=1}^{n} X_i X_j \mathbb{E}((R_i - \mathbb{E}(R_i))(R_j - \mathbb{E}(R_j))).\end{aligned}$$

We define the covariance of R_i and R_j via

$$\operatorname{Cov}(R_i, R_j) = \mathbb{E}((R_i - \overline{R}_i)(R_j - \overline{R}_j)). \tag{1.3}$$

(Recall $\overline{R}_i = \mathbb{E}(R_i)$.)

So

$$\operatorname{Var}\left(\sum_{i=1}^{n} X_i R_i\right) = \sum_{i,j=1}^{n} X_i \operatorname{Cov}(R_i, R_j) X_j. \tag{1.4}$$

If we let C be a matrix with entries

$$C_{ij} = \operatorname{Cov}(R_i, R_j),$$

we can rewrite the variance as

$$\mathrm{Var}\left(\sum_{i=1}^{n} X_i R_i\right) = x^t C x. \tag{1.5}$$

We call C the *covariance matrix* of the returns. It is clearly symmetric: that is $C_{ij} = C_{ji}$ for all i and j. In addition, the diagonal entries are simply the variances of the individual assets. Covariances of asset pairs may be negative. For example, if whenever one asset goes up, another tends to go down, they will be negatively correlated.

Note that to interpret (1.5) correctly, we regard x as a vector which is a matrix with one column and n rows. The matrix C is of size $n \times n$. The transpose of x is written x^{T}, and has one row and n columns, so

$$x^{\mathrm{T}} = (X_1, X_2, \ldots, X_n).$$

We are multiplying a $1 \times n$ matrix by a $n \times n$ matrix, and then by a $n \times 1$ matrix to get a 1×1 matrix, i.e. a number. Since this number is a variance, it is always greater than or equal to zero.

Definition 1.6 If C is a symmetric matrix and

$$x^{\mathrm{T}} C x \geq 0,$$

for all x, then C is said to be positive semi-definite. It is said to be positive definite if $x^{\mathrm{T}} C x > 0$, for $x \neq 0$.

So all covariance matrices are positive semi-definite, and furthermore it can be shown that any positive semi-definite matrix is the covariance matrix of some collection of random variables. (See [8].)

Note that the variance of an asset is the covariance of an asset with itself. It follows from (1.4) that the variance of a portfolio will be the sum of the variances of the individual assets if and only if the covariances between the pairs of the different asset pairs are all zero. That is, if and only if the assets are uncorrelated. We then have

$$\mathrm{Var}\left(\sum_{i=1}^{n} X_i R_i\right) = \sum_{i=1}^{n} X_i^2 \,\mathrm{Var}\, R_i.$$

Note that we can write the covariance of two asset returns R and S as

$$\mathrm{Cov}(R, S) = \sigma_R \sigma_S \rho_{RS},$$

where ρ_{RS} is referred to as the *correlation coefficient* and is defined in such a way as to make this statement true. Assets will have zero correlation if and only if they have zero covariance.

One condition that will lead to zero correlation is the much stronger condition of independence. If two random variables, C, D, are independent then

$$\mathbb{E}(CD) = \mathbb{E}(C)\mathbb{E}(D),$$

which in turn implies that

$$\text{Cov}(C,D) = \mathbb{E}(C-\overline{C})\mathbb{E}(D-\overline{D}) = 0.$$

When considering independent assets we might ask what happens if we take a large number of independent assets and invest the same fraction of our wealth in each? That is, suppose we take n assets and put $1/n$ into each asset. We have

$$\text{Var}\left(\sum_{i=1}^{n} \frac{1}{n} R_i\right) = \frac{1}{n^2} \sum_{i=1}^{n} \text{Var}(R_i).$$

If we assume that $\text{Var}(R_i) \leq C$ for some C for all i then we have

$$\text{Var}\left(\sum_{i=1}^{n} \frac{1}{n} R_i\right) \leq \frac{C}{n},$$

as n goes to infinity the variance will go to zero. This says that given a great enough number of independent assets, we can achieve an arbitrarily small amount of portfolio risk. What happens then if we allow covariance to be non-zero? In this case, we get

$$\begin{aligned}\text{Var}\left(\frac{1}{n}\sum_{i=1}^{n} R_i\right) &= \frac{1}{n^2}\sum_{i=1}^{n} \text{Var}(R_i) + \frac{2}{n^2}\sum_{i<j} \text{Cov}(R_i, R_j) \\ &= \frac{1}{n}\overline{\text{Var}(R_i)} + \frac{n-1}{n}\overline{\text{Cov}(R_i, R_j)}.\end{aligned}$$

Here we have used the fact that

$$1 + 2 + 3 + \cdots n - 1 = \frac{n(n-1)}{2},$$

so the number of elements in the sum $\sum_{i<j}$ is $n(n-1)/2$. Letting n tend to infinity, the variance converges to

$$\overline{\text{Cov}(R_i, R_j)}.$$

Thus by taking equal proportions of a large number of assets, we obtain a portfolio whose variance is the average covariance of the assets in the pool. This tells us a very important fact that the background covariance in a pool of assets will have an effect on how much risk we can diversify away. This suggests that we want to invest in as large a class of assets as possible.

1.5 Other risk measures

Variance can be criticised for penalising upside volatility as well as downside volatility. We generally only care about our possibility of loss, not our possibility of a windfall gain. With this in mind, we can define the *semi-variance* of a variable X via

$$\mathbb{E}((X - \mathbb{E}(X))^2 I_{X<\mu}),$$

where $I_{X<\mu}$ equals 1 for $X < \mu$ and 0 otherwise. Markowitz therefore devoted a chapter of his book, [11], to the problem of portfolio selection using semi-variance rather than variance. Here, however, we will tend to focus on cases where X is reasonably symmetric. It follows then that the semi-variance will not give much beyond the variance and so we will not study it further.

Another measure of risk, highly popular with financial institutions and regulators alike, is Value-At-Risk (VAR). The idea here is to define a maximum loss to be tolerated at a given level of probability over a given time horizon. We will discuss this measure in some detail, both in terms of its uses and shortcomings in Chapter 12.

1.6 Review

By the end of this chapter, the reader should be able to answer the following theoretical questions.

1. What is the objective of modern portfolio theory?
2. How is return defined in MPT?
3. How is expected return defined?
4. How do we maximise return if there are no risk constraints?
5. Derive the formula for the variance of returns of a portfolio.
6. What is a covariance matrix?
7. What special properties does a covariance matrix have?
8. Derive the formula for the variance of return of a large pool of correlated assets.
9. Define semi-variance.

1.7 Problems

Question 1.1 An asset has the following distribution of returns. Compute the mean return, standard deviation of returns, and semi-variance of returns.

Probability	Return
0.1	-3
0.1	-2
0.2	-1
0.2	0
0.2	1
0.1	2
0.1	10

Question 1.2 Suppose x and y are vectors. Let

$$\langle x, y \rangle = \sum_{i=1}^{n} x_i y_i.$$

For some fixed y, let $f(x) = \langle x, y \rangle$. Compute $\frac{\partial f}{\partial x_i}$.

Question 1.3 Let A be a square matrix with n rows. Let A^{T} denotes its transpose. Let u and v be vectors with n entries. Show

$$\langle Au, v \rangle = \langle u, A^{\mathrm{T}} v \rangle = u^{\mathrm{T}} A^{\mathrm{T}} v.$$

Question 1.4 Let A be a real symmetric square matrix (so $A = A^{\mathrm{T}}$), and let

$$g(x) = x^{\mathrm{T}} A x.$$

Compute

$$\frac{\partial g}{\partial x_k}.$$

Question 1.5 Assets A, B and C have expected returns of 8, 10, and 12. Their standard deviations of returns are 12, 10, and 8. The pairwise correlations are all 0.5. Find the variances and expected returns of the following portfolios:

- equal weights of all assets;
- 0.5 units of A, 0.25 units of B, and 0.25 units of C.

Question 1.6 Repeat the previous question with standard deviations 10, 10, and 10.

Question 1.7 Assume that the average standard deviation of return for an individual security is 7 and that the average correlation is 0.2. Estimate the standard deviation of returns of portfolios composed of 10 and 100 securities.

Question 1.8 Three assets, X_1, X_2, and X_3, have expected returns 10, 12, and 14 and covariance matrix of returns

$$\begin{pmatrix} 6 & 1 & 0 \\ 1 & 5 & 1 \\ 0 & 1 & 4 \end{pmatrix}.$$

Portfolio A is 0.5 units of X_1 and 0.5 units of X_2. Portfolio B is 0.5 units of X_2 and 0.5 units of X_3. Find the expected returns of A and B and their correlation of returns. (This is easier if you use Question 1.15.)

Question 1.9 What happens to the variance of a portfolio consisting of a large number of assets with returns that are bounded in variance in each of the two following cases:

- the assets' returns are all independent of each other;
- the mean covariance of the assets is $x > 0$.

Question 1.10 Suppose C_0 and C_1 are covariance matrices of random variables. For what values of θ is

$$C_\theta = (1-\theta)C_0 + \theta C_1$$

always a possible covariance matrix?

Question 1.11 Repeat the previous question replacing "covariance" with "correlation".

Question 1.12 An asset's returns are normally distributed with mean μ and variance σ^2. Compute its semi-variance.

Question 1.13 Suppose two assets' returns have continuous densities f and g both of which are symmetric about the mean return. Show that the variance divided by the semi-variance is the same in each case.

Question 1.14 If X and Y are random variables, and $\lambda, \mu \in \mathbb{R}$, find the covariance of λX and μY in terms of the covariance of X and Y.

Question 1.15 Suppose assets A_j, for $j = 1, \ldots, n$ have returns R_j and that these have a covariance matrix C. If portfolios X and Y have weights x and y, respectively, find an expression for the covariance of returns between X and Y.

Question 1.16 A square matrix has the value α on the diagonal and the value $\beta \geq 0$ off the diagonal. Give necessary and sufficient conditions on α and β for the matrix to be the covariance matrix of some collection of random variables. Justify your answer. (Condition B is necessary for property A if A implies B. Condition B is sufficient for property A if B implies A.)

2
Efficient portfolios: the two-asset case

2.1 Defining efficiency

In this chapter, we further develop our framework for portfolio analysis by using the mean and variance to characterise those portfolios that an investor will consider worthwhile. This is known as *mean–variance* analysis and whilst it has its flaws, it provides a good starting point for portfolio theory. The study of the investment problem from this perspective was initiated by Markowitz and his original book, [11], is still a good reference.

Our principal assumptions:

- Investors only care about the mean and variance of returns.
- Investors prefer higher means to lower means.
- Investors prefer lower variances to higher variances.
- We know the means, variances and covariances of returns for the assets we can invest in.

Put simply, all investors want more money and less risk.

Importantly, we assume that we know the mean and variance of the returns in the assets in which we can invest. However, these are **not** quantities which we can readily observe in the market. We will have to infer them from the data which we can observe and will return to this problem later in Chapter 5.

Definition 2.1 The set of all possible pairs of standard deviations and returns attainable from investing in a collection of assets is called the *opportunity set*.

Definition 2.2 A portfolio is *efficient* relative to a given opportunity set provided

- no other portfolio in that opportunity set has at least as much expected return and lower standard deviation, and
- no other portfolio in that opportunity set has higher return and standard deviation which is smaller or equal.

In other words, no other portfolio does at least as well as an efficient portfolio in terms of both risk and return, and better on one of these measures. An efficient portfolio therefore lies on the edge of the opportunity set. (If not on the edge, there is a direction that gives better return or risk.) We assume (for now) that investors only want efficient portfolios.

Note that we get the same set of portfolios if we replace standard deviation by variance in the definition. This follows since the mapping between variance and standard deviation is strictly increasing.

Definition 2.3 The subset of the opportunity set which is efficient is called the *efficient frontier*.

It is important to realise that efficiency is only defined relative to a set of investment opportunities. If we change the set of assets available to the investor, then the set of efficient portfolios changes too. In general, if we allow an extra asset then portfolios that were previously efficient will no longer be efficient. Similarly, if we throw away an asset both from the set of investment opportunities and from an efficient portfolio, then the portfolio containing the remaining assets may not be efficient.

2.2 Two-asset portfolios

To develop our understanding of these concepts, we consider the case where the opportunity set is defined by two assets, B and S, both of which are risky. We can think of B as a risky bond, and of S as a risky stock. We only have one variable, which is the fraction of investments put into B; everything remaining must be put into S. This means that the opportunity set will be one-dimensional and our analysis will now focus on understanding its shape under different circumstances.

We start with investment fractions X_S and X_B such that

$$X_S + X_B = 1.$$

We have, by using the relation between X_B and X_S,

$$\begin{aligned}\mathbb{E}(R_P) =& X_B\mathbb{E}(R_B) + (1 - X_B)\mathbb{E}(R_S),\\ =& X_B(\mathbb{E}(R_B) - \mathbb{E}(R_S)) + \mathbb{E}(R_S),\end{aligned}$$

and

$$\sigma_P^2 = X_B^2\sigma_B^2 + (1 - X_B)^2\sigma_S^2 + 2X_B(1 - X_B)\sigma_{BS}. \tag{2.1}$$

The expected return is linear in X_B whilst the variance is quadratic.

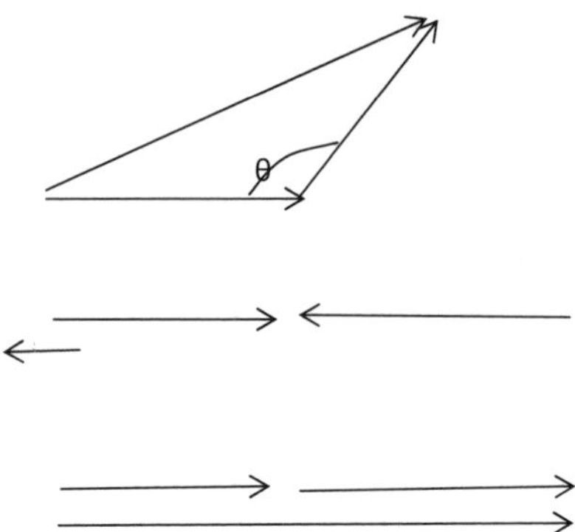

Figure 2.1 The addition of vectors with differing directions reflecting the correlations $\cos(\pi - \theta)$, -1 and 1.

Here P is the portfolio, R_P is its return, σ_P is the standard deviation of returns, and σ_{BS} is the covariance of returns of B and S.

We can solve for X_B in terms of the expected return, provided the two assets have different expected returns, and it is a linear function of the expected return:

$$X_B = \frac{\mathbb{E}(R_P) - \mathbb{E}(R_S)}{\mathbb{E}(R_B) - \mathbb{E}(R_S)}.$$

If we now substitute this back into the expression for variance, we obtain

$$\sigma_P^2 = \alpha \mathbb{E}(R_P)^2 + \beta \mathbb{E}(R_P) + \gamma$$

for some α, β and γ. The variance of P is therefore generally a parabola as a function of expected return for two-asset portfolios.

We can then differentiate in equation (2.1) to find the value of X_B that gives least variance. This will occur where the derivative is zero since that is the turning point of the parabola. We find that the critical point is at X_B^* given by

$$\begin{aligned} X_B^* &= \frac{\sigma_S^2 - \sigma_{BS}}{\sigma_B^2 + \sigma_S^2 - 2\sigma_{BS}}, \\ &= \frac{\sigma_S^2 - \sigma_B \sigma_S \rho_{BS}}{\sigma_B^2 + \sigma_S^2 - 2\sigma_B \sigma_S \rho_{BS}}. \end{aligned}$$

In the two-asset case, the portfolio of minimal variance will always be efficient, and this turns out to be true in general.

For those who like to interpret equations geometrically, equation (2.1) may be readily visualised in terms of vectors in the plane. That is, we can think of standard deviation as the length of a vector. Holding X units of an asset with return R, multiplies the length R of the vector by X. The correlation between the two assets' returns is then the cosine of the angle, θ, between the two corresponding vectors. The standard deviation of the two-asset portfolio is the length of the two vectors added together. We have for two vectors v and w with angle θ,

$$||v+w||^2 = ||v||^2 + ||w||^2 + 2||v||\,||w||\cos(\theta).$$

We illustrate this in Figure 2.1 This interpretation will be useful when attempting to understand how forming portfolios of two assets affects standard deviation.

2.2.1 The effect of correlation

Having done a little work on the general case for two assets, we now study some special cases in order to develop our intuition. In all of the following we take the parameters

$$\begin{aligned}\sigma_S =&0.15,\\ \sigma_B =&0.1,\\ \overline{R}_S =&0.06,\\ \overline{R}_B =&0.05.\end{aligned}$$

The all important number is ρ_{BS}, and it is this quantity which we will vary in the examples below.

Example 2.4 If the assets are perfectly correlated, i.e. $\rho_{BS} = 1$, there is no risk reduction arising from diversification, and we have that both standard deviation and return are linear functions of X_B. In terms of the geometric model, the two return vectors point in the same direction in the plane and $\cos\theta = 1$.

We have

$$\begin{aligned}\sigma_P^2 =&X_B^2\sigma_B^2 + (1-X_B)^2\sigma_S^2 + 2X_B(1-X_B)\sigma_B\sigma_S,\\ =&(X_B\sigma_B + (1-X_B)\sigma_S)^2\end{aligned}$$

which implies

$$\sigma_P = |X_B\sigma_B + (1-X_B)\sigma_S|.$$

The opportunity set will be a straight line in return/standard deviation space but will have a sharp turn when we pass through zero, since standard deviation is always non-negative. We graph this case in Figure 2.2. ◇

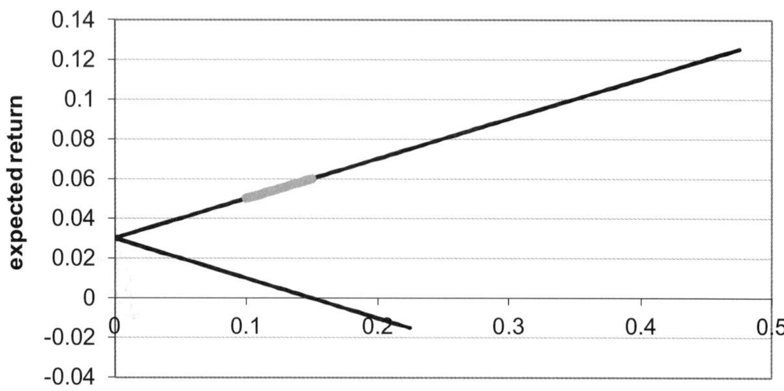

Figure 2.2 Expected return against standard deviation for two perfectly correlated assets for varying investment weights. The section with no short sales is highlighted.

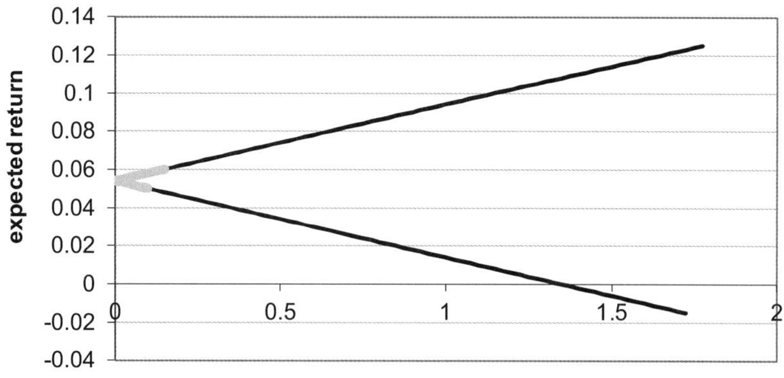

Figure 2.3 Expected return against standard deviation for two assets with perfect negative correlation. No short-selling part highlighted.

Example 2.5 Suppose next that we have $\rho_{BS} = -1$. There will be a point where the two pieces of risk cancel each other out and we obtain zero risk. Geometrically, the two vectors point in exactly opposite directions in the plane.

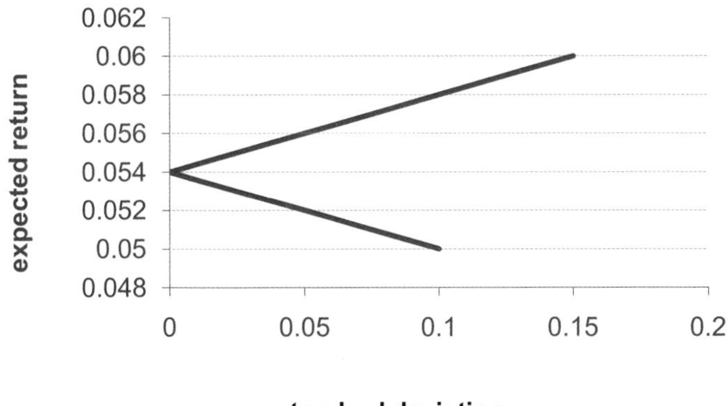

Figure 2.4 Expected return against standard deviation for two assets with perfect negative correlation. No short sales allowed.

We have

$$\begin{aligned}\sigma_P^2 =& X_B^2\sigma_B^2 + (1-X_B)^2\sigma_S^2 - 2X_B(1-X_B)\sigma_B\sigma_S,\\ =& (X_B\sigma_B - (1-X_B)\sigma_S)^2.\end{aligned}$$

This implies

$$\sigma_P = \left|X_B\sigma_B - (1-X_B)\sigma_S\right|.$$

Note the absolute value sign here: standard deviation can never be negative. The absolute value sign implies the graph of the opportunity set is piecewise linear: it is linear for X_B above and below the point where σ_B is zero but not across the point where it is zero, see Figure 2.3.

Note how similar the anti-correlated and perfectly correlated cases. This is because if X and Y have correlation one, then X and $-Y$ have correlation -1.

Under the assumption of no short-selling, the graph is truncated as shown in Figure 2.4. ◇

Example 2.6 We now consider the case where $\rho_{BS} = 0$; that is the asset returns are uncorrelated. This corresponds to taking the length of the sum of two orthogonal vectors in the plane (i.e. $\cos\theta = 0.$).

We have

$$\sigma_P^2 = X_B^2\sigma_B^2 + (1-X_B)^2\sigma_S^2.$$

This time we get a curve in mean/standard deviation space and at the mini-

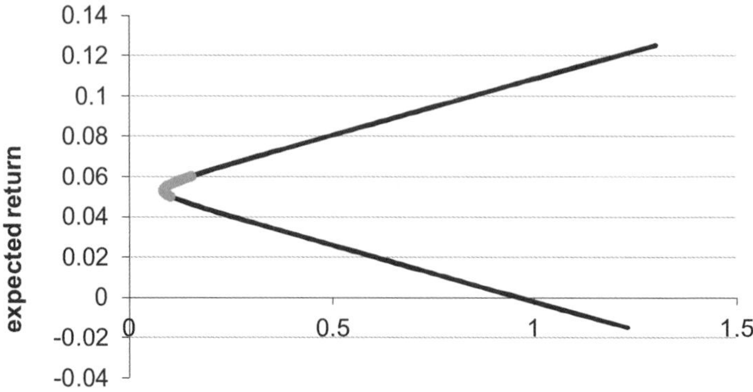

Figure 2.5 Expected return against standard deviation for two assets with zero correlation. No short-selling part highlighted.

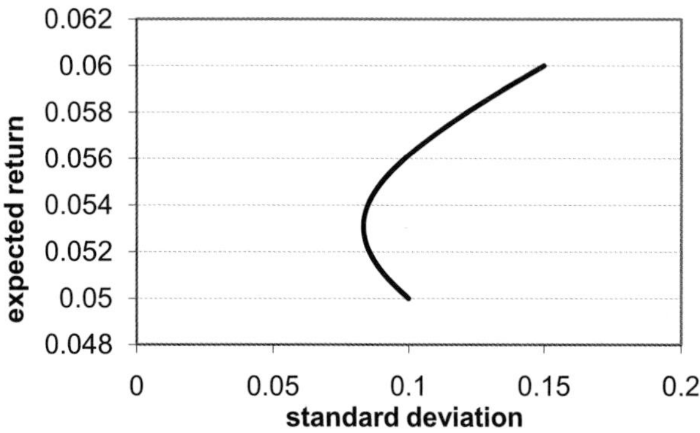

Figure 2.6 Expected return against standard deviation for two assets with zero correlation. No short sales allowed.

mum, we obtain

$$X_B^* = \frac{\sigma_S^2}{\sigma_B^2 + \sigma_S^2}.$$

This is illustrated in Figure 2.5; the portion not involving short-selling is highlighted. This portion is magnified in Figure 2.6. ◇

Having examined above perfect correlation, anti-correlation and zero corre-

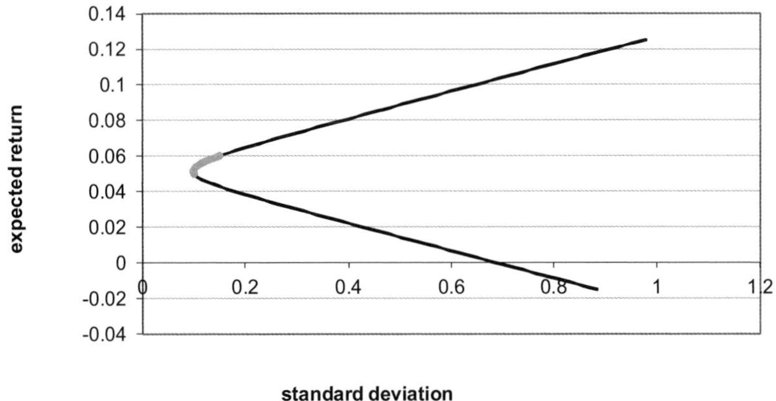

Figure 2.7 Expected return against standard deviation for two assets with 0.5 correlation. No short-selling part highlighted.

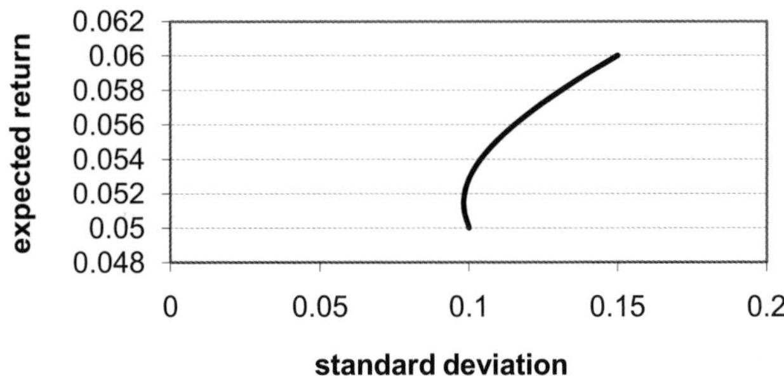

Figure 2.8 Expected return against standard deviation for two assets with 0.5 correlation. No short sales allowed.

lation, we next turn to the case where two assets are correlated but not perfectly so. As one might suspect, we will obtain a curve somewhere between the perfect correlation case and the uncorrelated case.

Example 2.7 Suppose $\rho_{BS} = 0.5$ and all other assumptions remain as in the earlier examples. Where short-selling is allowed we obtain a curve as shown in Figure 2.7. With no short-selling, the curve is as in Figure 2.8. We can further develop our intuition by varying the correlation as shown in Figure 2.9. ◇

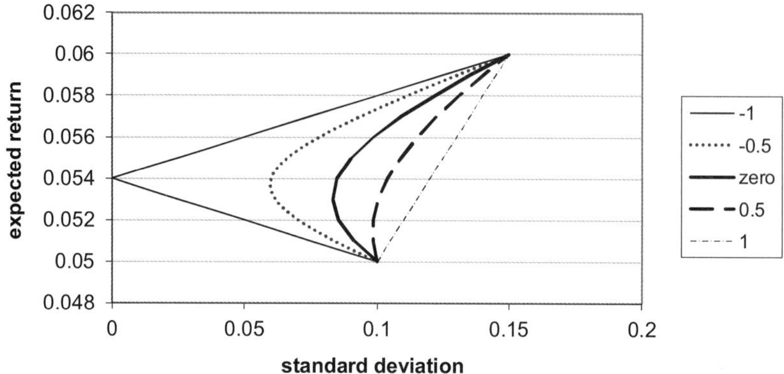

Figure 2.9 Expected return against standard deviation for two assets with varying correlations. No short sales allowed.

2.2.2 Classifying the curves

There is an area of mathematics that studies the zero sets of quadratic functions of two variables, namely *conic sections*: they are so called because they can be realised as the intersection of a cone with a plane.

With two assets, the opportunity set is a hyperbola as return is linear in X_B which in turn means that X_B is linear in return, and we have

$$\sigma_P^2 = X_B^2\sigma_B^2 + (1-X_B)^2\sigma_S^2 + 2X_B(1-X_B)\sigma_{BS}.$$

It can then be shown that

$$\sigma_P^2 = \alpha\overline{R}_P^2 + \text{ lower order terms,}$$

with σ_P constrained to be non-negative.

Note that if there are no lower-order terms, then one obtains

$$\sigma_P = \pm\overline{R}_P,$$

which describes two intersecting lines. That is, two intersecting lines is a special case of a hyperbola corresponding to the cases of perfect correlation and perfect anti-correlation.

Alternatively, if we work in mean/variance space, then the equation is

$$V = \alpha\overline{R}_P^2 + \text{ lower order terms,}$$

and we have a parabola.

A function is said to be convex if the chord between any two points lies above the graph of the function. A function is said to be concave if the chord between two points lies below the graph. Hyperbolas are concave above the turning point. Straight lines are trivially both convex and concave.

Hence, we can see that the efficient frontier of two assets is concave, noting that the area below the turning point is not concave and not efficient.

2.3 Review

By the end of this chapter, the reader should be able to answer the following theoretical questions.

1. What are the assumptions of mean–variance portfolio theory?
2. What does it mean for an asset to be mean–variance efficient?
3. Define the opportunity set and efficient frontier in mean-variance analysis? How do they relate to each other?
4. If a portfolio is efficient and we add in a new asset to invest in will the original portfolio remain efficient?
5. If a portfolio is efficient and we discard one of its elements from the set of possible assets, will the part of the portfolio excluding this element always be efficient?
6. Derive expressions for the variance and expected returns of a portfolio of two assets in terms of the amount invested in the first, their expected returns, their variances and covariance.
7. Describe the shape of the graph of expected return of two asset portfolios as a function of the investment fraction.
8. Describe the shape of the graph of standard deviation of two asset portfolios as a function of the expected return.
9. Derive an expression for the composition of the minimal variance portfolio for possible investment assets.
10. Will the minimal variance portfolio always be efficient?
11. What does it mean for a graph to be concave? Convex? Is the efficient frontier convex and/or concave in general?

2.4 Problems

Question 2.1 Rank the following assets as far as possible using mean–variance efficiency if they have the following returns:

(1) a normal distribution X with mean 2 and variance 3;

(2) e^Z, with Z standard normal;
(3) a uniform distribution on the interval $[0,4]$;
(4) a uniform distribution on the interval $[1,5]$;
(5) a uniform distribution on the interval $[0,5]$.

Question 2.2 Random variables X_1, X_2 and X_3 are independent with means $1,2$ and 3; they have variances $3,2$ and 1. Companies A_j have returns R_j as follows:

$$R_1 = 2X_1 + X_3;$$
$$R_2 = X_1 + X_2 + X_3;$$
$$R_3 = 2X_2 + X_3.$$

Rank these investments using mean–variance analysis as far as possible.

Question 2.3 Two assets A and B have expected returns of 10 and 5, respectively. The standard deviations of their returns are 8 and 6, respectively. If they are independent, what is the expected return and standard deviation of the minimal variance portfolio? What is the maximal expected return that can be achieved if short-selling is not allowed?

Question 2.4 Two assets C and D have expected returns of 5 and 10, respectively. The standard deviations of their returns are 10 and 5, respectively. For each of the correlations -1, 0, 0.5 and 1, what is the expected return and standard deviation of the minimal variance portfolio? What is the maximal expected return that can be achieved if short-selling is not allowed?

Question 2.5 Given two risky assets, sketch the possible geometries of the opportunity set in expected return/standard deviation space in each of the following cases if short-selling is allowed:

- the returns have correlation -1;
- the returns have correlation 0;
- the returns have correlation 1.

Question 2.6 Given two risky assets, sketch the possible geometries of the efficient frontier in expected return/standard deviation space in each of the following cases if short-selling is not allowed:

- the returns have correlation -1;
- the returns have correlation 0;
- the returns have correlation 1.

Sketch a typical case with all three on one graph.

Question 2.7 You are in Australia where there are only two assets: oil and stocks. Past data analysis reveals that oil has average return 10 and standard deviation 25, whereas stocks have 12 and 22. The correlation is 0.2.

1. Which asset is better for a mean–variance investor?
2. A portfolio is half oil and half stocks, find its expected return and standard deviation.
3. Find the minimal variance portfolio.
4. It is rumoured that a drilling company is going to vary the amount of oil produced depending on stock prices in Australia by producing less oil when the stock market is up and more oil when the stock market is down. What effect will this have on portfolios composed of oil and stocks? Explain.

3
Portfolios with a risk-free asset

3.1 The risk-free asset

In this chapter we explore efficiency for portfolios containing a theoretical risk-free asset; that is, one whose return is known in advance. We will show that the addition of a risk-free asset to a portfolio of risky assets has an interesting effect on the geometry of the opportunity set; it reduces to a straight line. We go on to use this fact to show that any efficient portfolio (for a given set of investment opportunities) can be represented as a combination of a risk-free asset and a fixed portfolio of risky assets. Finding this portfolio, referred to as the "tangent portfolio," allows us to sweep out the efficient frontier, and in this chapter we illustrate this approach for the case of two risky assets. We generalise this approach to the multi-asset case in Chapter 4.

We start by defining a risk-free asset in the expected way.

Definition 3.1 An asset whose return is known in advance is said to be risk-free. An asset, C, is risk-free if and only if:

- the variance of returns is zero;
- the standard deviation of returns is zero.

All risk-free assets have the same return. We can see this easily by supposing there to be a second asset B, also risk-free but with a different return. Then everyone would sell the risk-free asset with lower return and buy the one with higher return until the returns eventually agreed. For our purposes we may therefore assume that there is a single risk-free asset.

Suppose next that our portfolio P consists of $1-y$ units of the risk-free asset, C, with return R_f, and y units of some risky asset (or portfolio of assets) A with return R_A; then the expected return of P is

$$\overline{R}_P = (1-y)R_f + y\overline{R}_A,$$

and since R_f is riskless, we have

$$\begin{aligned}\operatorname{Var}(R_p) =&\operatorname{Var}(yR_A),\\ =&y^2\operatorname{Var}(R_A).\end{aligned}$$

Taking square roots, we have

$$\sigma_P = |y|\sigma_A.$$

Note if $y = 0$, then we get the risk-less asset back. If $y = 1$, we get A back. If we restrict to $y \geq 0$, we have that

$$y = \frac{\sigma_P}{\sigma_A}.$$

That is, the investment fraction in A is the ratio of the standard deviations of the portfolio and the risky asset. This implies

$$\overline{R}_P = \left(1 - \frac{\sigma_P}{\sigma_A}\right) R_f + \frac{\sigma_P}{\sigma_A}\overline{R}_A,$$

and hence

$$\overline{R}_p = R_f + \frac{\overline{R}_A - R_f}{\sigma_A}\sigma_P.$$

The expected return is a straight line in return/standard deviation space, with slope

$$\frac{\overline{R}_A - R_f}{\sigma_A}.$$

We have that the combination of risk-free asset and a fixed risky portfolio lead to a straight-line in return/standard devation space.

If we let the risk be zero, we get the risk-free asset. If we let the risk be equal to that of A, we get A. In conclusion, we have a straight line through the risk-free asset and A in return/standard deviation space. Importantly, this is true for any A, whether a single asset or a portfolio. This straight line is often called the *investment line.* We can interpret it as follows. Between the two points, we are dividing our portfolio into risk-free assets and risky-assets. Above the risky point, we are short-selling the risk-free asset, and putting even more money into the risky assets.

We know that the market demands a higher return for a risky asset over the risk-free return. The slope of the investment line tells us just how much extra return is available. To see this, note that we have

$$\overline{R}_P = R_f + \frac{\overline{R}_A - R_f}{\sigma_A}\sigma_P.$$

This says that for each extra unit of risk expressed as standard deviation that we take, we get

$$\lambda = \frac{\overline{R}_A - R_f}{\sigma_A}$$

extra units of expected return.

We can regard λ as the *market price of the risk* on A. We illustrate this in the following example.

Example 3.2 We have

$$\begin{aligned} R_f &= 3, \\ \sigma_A &= 12, \\ \overline{R}_A &= 12. \end{aligned}$$

The market price of risk for A is

$$\frac{12-3}{12} = \frac{3}{4}.$$

The reader is now invited to calculate the market price of risk for B where $\overline{R}_B = 15$, and $\sigma_B = 6$. ◇

3.2 Efficiency with a risk-free asset

Now suppose we have a portfolio containing a risk-free asset, C, and a collection of many risky assets. Suppose furthermore that this portfolio E is efficient. We can write E in the form

$$X_f C + (1 - X_f)A,$$

for some portfolio of risky assets, A, and $(1 - X_f) > 0$. Note that the latter assumption is reasonable since if this were not the case, we would replace the portfolio A with $-A$.

It turns out that if we now discard the risk-free asset, then A is itself efficient amongst the set of risky assets:

Theorem 3.3 *If E is efficient then the portfolio A consisting of the risky assets in E is efficient relative to investing solely in risky assets.*

The intuitive reason that this result holds is that R_f has no variance or co-variance so no diversification effects can come into play. However, if we added or subtracted a risky asset, a similar result would not hold. We demonstrate this more rigorously below.

Proof Before proceeding to the proof, observe its structure.

- We use proof by contra-positive.
- We show that if A is not efficient then neither is E.
- This is logically equivalent to the statement that if E is efficient then A is efficient.

Note that in the proof C is riskless so

$$\mathbb{E}(R_C) = R_C = R_f.$$

If A is not efficient, then either there exists

(1) a portfolio of risky assets B with higher return and the same or lower variance; or
(2) a portfolio D with the same return and lower variance.

In case (1),

$$\begin{aligned}\mathbb{E}(X_f R_C + (1-X_f)R_B) &= X_f R_f + (1-X_f)\mathbb{E}(R_B),\\ &> X_f R_f + (1-X_f)\mathbb{E}(R_A),\end{aligned}$$

so we get higher return; and

$$\begin{aligned}\mathrm{Var}(X_f R_C + (1-X_f)R_B) &= \mathrm{Var}((1-X_f)R_B),\\ &= (1-X_f)^2\mathrm{Var}(R_B),\\ &\leq (1-X_f)^2\mathrm{Var}(R_A).\end{aligned}$$

So in case (1), we have that $X_f C + (1-X_f)A$ is not efficient.

In case (2),

$$\begin{aligned}\mathbb{E}(X_f R_C + (1-X_f)R_D) &= X_f R_f + (1-X_f)\mathbb{E}(R_D),\\ &= X_f R_f + (1-X_f)\mathbb{E}(R_A),\end{aligned}$$

so we get the same return; and

$$\begin{aligned}\mathrm{Var}(X_f R_C + (1-X_f)R_D) &= \mathrm{Var}((1-X_f)R_D),\\ &= (1-X_f)^2\mathrm{Var}(R_D),\\ &< (1-X_f)^2\mathrm{Var}(R_A).\end{aligned}$$

So in case (2), we have that $X_f C + (1-X_f)A$ is not efficient.

So if A is not efficient then neither is $X_f C + (1-X_f)A$. Turning this around, if $X_f C + (1-X_f)A$ is efficient then so is A. □

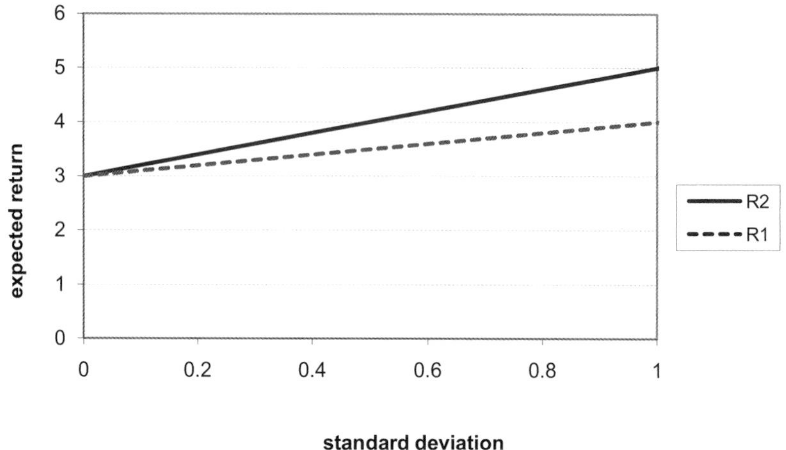

Figure 3.1 Two different investment lines with the same riskless asset.

We showed that if we took a portfolio of risky assets, then, by combining with the risk-free asset, the opportunity set is a straight line (i.e. the investment line). Hence, if we have an efficient portfolio

$$P = X_f C + (1 - X_f)A,$$

then the entire line through the points $(0, R_f)$ and $(\sigma_A, \overline{R}_A)$, including P, can be found by investing in proportions of A and C. In fact, this entire line is efficient.

Theorem 3.4 *If there is a risk-free asset, all efficient portfolios lie on a straight line in standard deviation/expected return space.*

Proof Suppose we were to have an efficient portfolio that was not on this line. Then the entire line through it and $(0, R_f)$ would also be in the opportunity set. Two straight lines through R_f will either be the same, or one will be below the other for all $\sigma_P > 0$, which means that none of the portfolios on it are efficient.

So the two lines must be the same. □

In Figure 3.1 we show two investment lines which pass through the point $(0, 3)$ corresponding to investing all monies into the risk-free asset.

We have shown that if C is riskless and

$$X_C C + (1 - X_C)A$$

is efficient for *some* X_C then it is efficient for *all* X_C. In particular, the portfolio consisting only of C is efficient, and the portfolio consisting only of A is efficient.

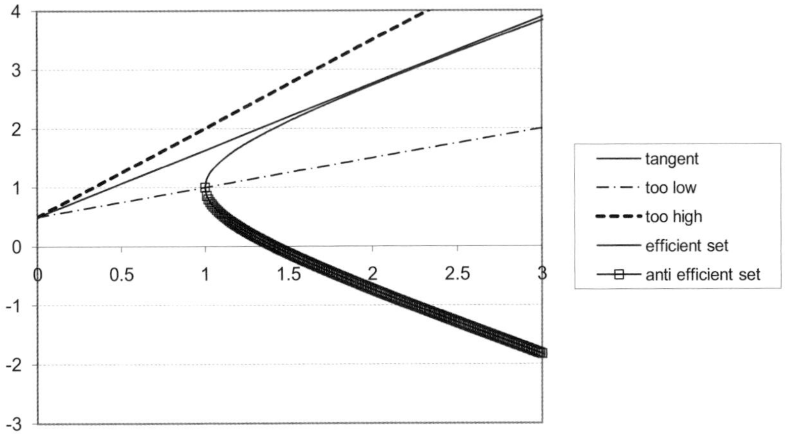

Figure 3.2 Investment lines and the opportunity set for two riskless assets.

3.3 Tangent portfolios

All efficient portfolios can be obtained as a mixture of a single portfolio of risky assets A and the risk-free asset C. It follows that mean–variance investors with the same investment situation will hold the same portfolio of risky assets A but may hold differing proportions of A.

Note the crucial point here is that the investors have to have everything in common such as views on means and variances, and taxes. However, they are not required to have the same risk preferences. (A lower risk investor will simply hold a smaller proportion of A and a correspondingly higher proportion of C.) We have shown

Theorem 3.5 (Tobin's separation theorem) *Two mean–variance investors facing the same investment situation will hold the same portfolio of risky assets.*

Note that we have not proven that A is unique, and we have not made enough assumptions for this to be true. However, it will be generally be true unless it is possible to make a risk-free asset from a combination of risky assets.

The portfolio A is often called the *tangent portfolio*.

We next explore this concept of tangency with two risky assets. That is, suppose we have two risky assets with imperfect correlation and one risk-free asset. We know:

- the efficient set without the risk-free asset is a hyperbola;
- there will be a portfolio, A, which is efficient with and without the risk-free asset; and
- the efficient set will be a straight line.

A line can intersect a hyperbola in three ways:

- in no points;
- in one point, and is tangent;
- in one or two points and is not tangent.

The line through the risk-free asset that is *tangent* to the hyperbola will be the efficient line for investments including it. To see this, consider the alternative cases:

- if it intersects but is not tangent then it will still intersect it if moved up a little (thereby increasing return) so it is not efficient;
- if it does not meet the hyperbola then it is not in the opportunity set, since any point must be on the straight line from the risk-free asset to a portfolio consisting purely of risky assets.

We illustrate these cases in Figure 3.2.

To summarise, for two imperfectly correlated risky assets and one risk-free asset:

- the efficient set is a straight line;
- this straight line passes through the risk-free asset;
- this straight line is tangent to the hyperbola which is the efficient set for the two risky assets;
- the efficient set of portfolios consists of linear combinations of the risk-free asset and the portfolio where the straight line is tangent to the hyperbola.

The portfolio at the intersection of the efficient line and the hyperbola is called the *tangent portfolio*.

3.4 Examples

We illustrate these findings with some examples. In the first, suppose we can invest in a Canadian bond fund, B, and an index tracker on the stock market, S. We want to find the efficient frontier.

Bonds offer lower returns but also lower volatility than stocks. The two

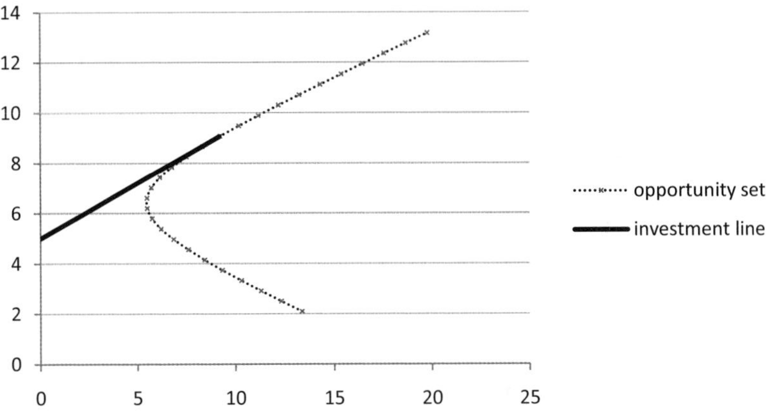

Figure 3.3 Opportunity set and part of the investment line for Canadian bond fund and index tracker example.

funds are correlated since both are affected by the overall economy. We are given

$$\overline{R}_S = 10.3,$$
$$\sigma_S = 12.2,$$
$$\rho_{S,B} = 0.34,$$
$$\overline{R}_B = 6.2,$$
$$\sigma_B = 5.5.$$

Using the formula for the minimal variance portfolio, we compute

$$X_B = 0.944.$$

We hold virtually no stocks for minimal variance. Substituting, we have that the minimal standard deviation is 5.46 which is slightly better than for bonds alone.

Suppose next that we add in a risk-less investment. We let the risk-free rate be 5. We can determine the tangent point graphically (or analytically for those who prefer). It has

$$\overline{R}_T = 9.06,$$
$$\sigma_T = 9.21:$$

see Figure 3.3. Note that the investment line would extend beyond the tangent point; we have stopped there to make it clearer where it is.

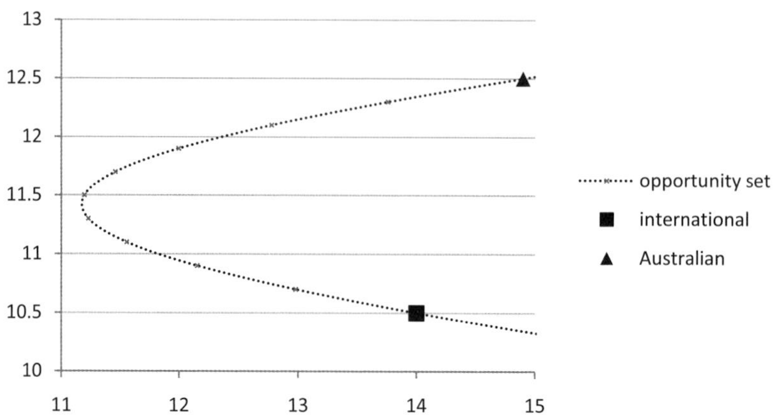

Figure 3.4 Opportunity set for example of an Australian stock fund and an international fund.

The tangent line has slope

$$\frac{9.06 - 5}{9.21} = 0.44,$$

and so its equation is

$$\overline{R}_P = 5 + 0.44\sigma_P.$$

To get the tangent portfolio, we know

$$\overline{R}_P = X_S\overline{R}_S + (1 - X_S)\overline{R}_B,$$

and so

$$9.06 = X_S(10.3) + (1 - X_S)6.2,$$

which implies

$$X_S = 0.697, \text{ and } X_B = 0.303.$$

Suppose for comparison, we now consider a mix of an Australian stock fund and an international fund. Certainly, we can expect lower correlation, i.e., more diversification, since the latter will be less reliant on local performance.

Our parameters are

$$\begin{aligned}
\overline{R}_S =& 12.2,\\
\sigma_S =& 14.5,\\
\rho_{S,I} =& 0.2,\\
\overline{R}_I =& 10.7,\\
\sigma_I =& 14.3.
\end{aligned}$$

This portfolio will differ from the earlier stock–bond example in the following ways:

- lower correlation due to the international component;
- higher return on second asset since it is a stock portfolio;
- higher standard deviation on second asset since it is a stock portfolio.

This time we obtain for our minimal variance portfolio

$$\begin{aligned}
X_B =& \frac{14.5^2 - 0.2(14.5)(14.3)}{14.5^2 + 14.3^2 - 2(0.2)(14.5)(14.3)},\\
=& 0.491.
\end{aligned}$$

We have roughly equal proportions of the two funds.

The minimal standard deviation is 11.15 and the expected return is 11.44. It follows that the international fund is not efficient; it has both a lower return and a higher standard deviation than the minimal variance portfolio. We illustrate the opportunity set together with the location of the two constituent funds in Figure 3.4.

Next suppose we again add in a risk-less investment with rate 5. We can again determine the tangent point (graphically or analytically.) It has

$$\begin{aligned}
\overline{R}_T =& 11.57,\\
\sigma_T =& 11.26.
\end{aligned}$$

We get slope

$$\frac{11.56 - 5}{11.26} = 0.583,$$

and the efficient line is

$$\overline{R}_P = 5 + 0.583\sigma_P.$$

We can solve, as before, to get

$$X_S = 0.579.$$

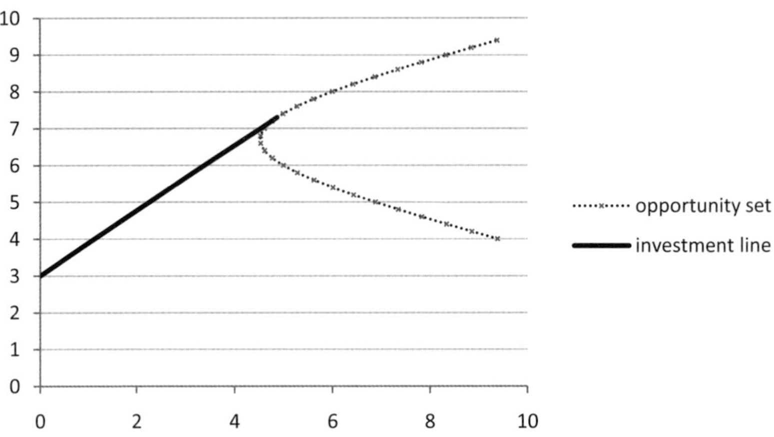

Figure 3.5 The investment line terminates at the tangent portfolio if borrowing is not allowed.

3.5 Borrowing restrictions

Riskless borrowing is only possible if you are a government with an impeccable credit rating (of which there are now fewer than once thought). It is therefore generally not possible to short a riskless asset. To see this we observe that shorting the riskless asset is the same as borrowing at the risk-free rate. The buyer of the asset is taking the risk that you will not pay back the money and will demand an appropriate risk premium. It is therefore reasonable to impose restrictions on the riskless asset.

We recall that the tangent portfolio corresponds to having all monies invested in the stock (or risky assets). If we restrict to no borrowing then the investment line will terminate at the tangent portfolio.

We can therefore invest in all assets on the line between the riskless asset and the tangent portfolio. The part of the hyperbola continuing on above the tangent portfolio will also be efficient. We illustrate this in Figure 3.5 where the investment line now stops at the tangent portfolio.

As an investor, when we buy the riskless bond we are lending at the riskless rate to an essentially riskless counterparty: the government. When we borrow, we are charged a risk premium by our financier. It follows that if borrowing is permitted, then the borrowing rate should be higher than the lending rate. The efficient frontier has a corresponding change of slope at the tangent point.

The efficient frontier will be in three pieces:

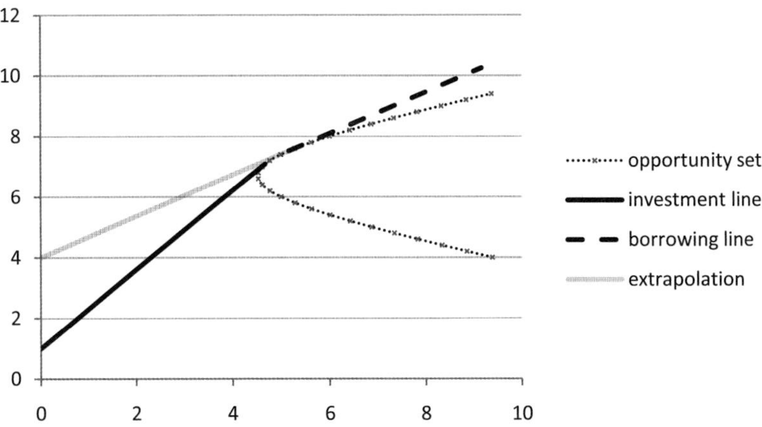

Figure 3.6 Two-asset opportunity set hyperbola together with efficient investment and borrowing lines.

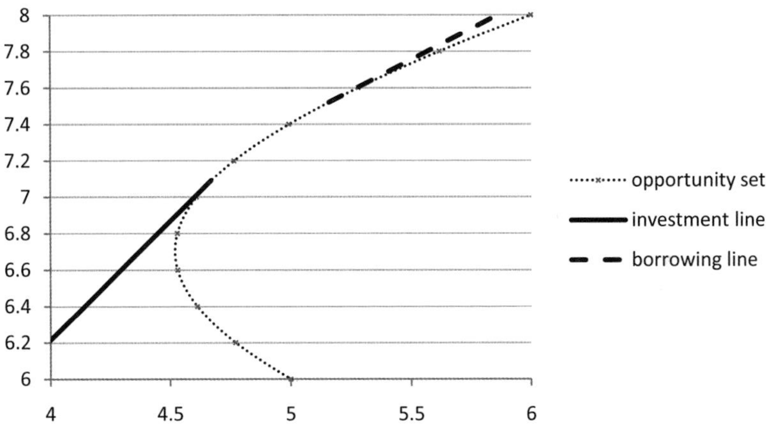

Figure 3.7 Two-asset opportunity set hyperbola together with efficient investment and borrowing lines. Section close to tangent portfolios magnified.

(1) the straight line to the tangent portfolio with riskless asset with the low lending rate;
(2) the section of the hyperbola between the tangent portfolio with the low lending rate and the tangent portfolio with high borrowing rate;
(3) the investment line beyond the tangent portfolio for the high borrowing rate.

This is shown in Figure 3.6 and is blown up in Figure 3.7.

3.6 Review

By the end of this chapter, the reader should be able to answer the following theoretical questions.

1. What does Tobin's separation theorem say?
2. Show that if a portfolio consisting of X units of A and $(1-X)$ units of the riskless assets is efficient then A is efficient if we cannot invest in the riskless asset.
3. Show that if a portfolio consisting of X units of A and $(1-X)$ units of the riskless assets is efficient then A by itself is efficient if we can invest in the riskless asset.
4. We have two risky assets and one riskless asset, sketch the efficient frontier with and without the riskless asset on the same graph.
5. Describe the shape of the efficient frontier in the space of weights when we have a riskless asset.
6. How would borrowing and lending rates vary for most investors?
7. What is the shape of the efficient frontier with two risky assets and one riskless asset, if no borrowing is possible?
8. What is the shape of the efficient frontier with two risky assets and one riskless asset, with different borrowing and lending rates?
9. Describe how to find the weights in all efficient portfolios with two risky assets and different borrowing and lending rates.

3.7 Problems

Question 3.1 In each of the following cases, find the market price of risk:

- $R_f = 8,\ \overline{R}_A = 12,\ \sigma_A = 10;$
- $R_f = 0,\ \overline{R}_A = 1,\ \sigma_A = 1;$
- $R_f = 5,\ \overline{R}_A = 15,\ \sigma_A = 5;$
- $R_f = 8,\ \overline{R}_A = 6,\ \sigma_A = 3.$

Which of these cases are realistic?

Question 3.2 In each of the following cases, if A is the tangent portfolio, find an efficient portfolio B with the specified standard deviation, and state its expected return:

- $R_f = 6,\ \overline{R}_A = 12,\ \sigma_A = 10,\ \sigma_B = 20;$
- $R_f = 5,\ \overline{R}_A = 10,\ \sigma_A = 5,\ \sigma_B = 10;$
- $R_f = 2,\ \overline{R}_A = 6,\ \sigma_A = 5,\ \sigma_B = 1.$

Question 3.3 In each of the following cases, if A is the tangent portfolio, find an efficient portfolio B with the specified expected return and state its standard deviation:

- $R_f = 6,\ \overline{R}_A = 12,\ \sigma_A = 10,\ \overline{R}_B = 8;$
- $R_f = 5,\ \overline{R}_A = 10, \sigma_A = 5,\ \overline{R}_B = 15;$
- $R_f = 2,\ \overline{R}_A = 6,\ \sigma_A = 5,\ \overline{R}_B = 2.$

Question 3.4 If A is the tangent portfolio, is it possible to find an efficient portfolio with the specified characteristics in each of the following cases:

- $R_f = 4,\ \overline{R}_A = 12, \sigma_A = 16,\ \overline{R}_B \geq 8,\ \sigma_B \leq 10;$
- $R_f = 3,\ \overline{R}_A = 10,\ \sigma_A = 5,\ \overline{R}_B \geq 15,\ \sigma_B \leq 20;$
- $R_f = 2,\ \overline{R}_A = 5,\ \sigma_A = 7,\ \overline{R}_B \geq 10,\ \sigma_B \leq 50.$

Question 3.5 Suppose the risk-free rate is 3. Suppose there are four other assets with expected returns 4, 5, 6, 7. An efficient portfolio is known to contain 0.2 units of the risk-free asset, and 0.2, 0.25, 0.15, and 0.2 units of the other assets, respectively. Find the composition of:

- the tangent portfolio;
- an efficient portfolio with return 10.

What is the expected return of the tangent portfolio?

Question 3.6 Suppose the risk-free rate is 3. Suppose there are 4 other assets with expected returns 6, 5, 6, 7. All the risky assets have standard deviation 5 and pairwise correlation 0.2. The tangent portfolio is known to contain $\frac{1}{4}$, $\frac{1}{12}$, $\frac{1}{4}$ and $\frac{5}{12}$ units of the four assets. Find efficient portfolios with standard deviations 1 and 10.

Question 3.7 Suppose the risk-free rate is 5. Suppose there are 4 other assets with expected returns 10, 10, 15 and 20. All the risky assets have standard deviation 5 and are pairwise independent. The tangent portfolio is known to contain $\frac{1}{7}$, $\frac{1}{7}$, $\frac{2}{7}$ and $\frac{3}{7}$ units of the four assets. What is the maximal expected return that can be achieved with standard deviation at most 5?

Question 3.8 An investor can lend at 1 and borrow at 5. We have two assets with expected returns 10 and 12, standard deviations 11 and 13 and correlation zero. Find the minimal variance portfolio excluding the riskless asset. From a graph, it is known that the two tangent portfolios have (standard deviation, return), $(8.439, 10.933)$, and $(8.516, 11.001)$. Find

- the investment weights (i.e. the proportion of assets in the portfolio) for each of the two tangent portfolios;
- the investment weights and standard deviation for efficient portfolios with expected returns 3, 6, 10.95 and 12;
- the investment weights and expected returns for efficient portfolios with standard deviations 3, 6, 8.47, 10 and 12.

Question 3.9 An investor can lend at 2 and borrow at 6. We have two assets with expected returns 9 and 12, standard deviations 12 and 13 and correlation 0.5. Find the minimal variance portfolio excluding the riskless asset. From a graph, it is known that the two tangent portfolios have (standard deviation, return), $(11.30, 11.07)$, and $(12.61, 11.83)$. Find

- the investment weights for each of the two tangent portfolios;
- the investment weights and standard deviation for efficient portfolios with expected returns 3, 6, 11.5 and 12;
- the investment weights and expected returns for efficient portfolios with standard deviations 3, 6, 12 and 15.

Question 3.10 An investor states that he wants to fully invest his portfolio of $1,000,000$ in two Australian stocks. He also states that he is able to borrow money at a rate of 8%, and he is a mean–variance investor. Describe the geometry of the space of possible investments in terms of:

- asset weightings;
- expected return versus standard deviation.

Given expected returns and the covariances of returns, how would you compute the set of possible weights?

Question 3.11 If A is the tangent portfolio for borrowing, is it possible to find an efficient portfolio with the specified characteristics in each of the following cases if we can borrow at the riskless rate plus 1?

- $R_f = 4$, $\overline{R}_A = 12$, $\sigma_A = 16$, $\overline{R}_B \geq 8$, $\sigma_B \leq 10$;
- $R_f = 3$, $\overline{R}_A = 10$, $\sigma_A = 5$, $\overline{R}_B \geq 15$, $\sigma_B \leq 20$;
- $R_f = 2$, $\overline{R}_A = 5$, $\sigma_A = 7$, $\overline{R}_B \geq 10$, $\sigma_B \leq 50$.

Question 3.12 If A is the tangent portfolio for borrowing, is it possible to find an efficient portfolio with the specified characteristics in each of the following cases if we can borrow at the riskless rate plus 5?

- $R_f = 4$, $\overline{R}_A = 12$, $\sigma_A = 16$, $\overline{R}_B \geq 8$, $\sigma_B \leq 10$;
- $R_f = 3$, $\overline{R}_A = 10$, $\sigma_A = 5$, $\overline{R}_B \geq 15$, $\sigma_B \leq 20$;
- $R_f = 2$, $\overline{R}_A = 5$, $\sigma_A = 7$, $\overline{R}_B \geq 10$, $\sigma_B \leq 50$.

4

Finding the efficient frontier – the multi-asset case

In this chapter we present the algorithm for finding the tangent portfolio and sweeping out the efficient frontier in the multi-asset case. For those who are interested in the mathematics supporting these mechanics we include an optional section at the end of the chapter.

4.1 Finding the tangent portfolio

Given a risk-free asset with return R_f and a risky portfolio A, we saw that the portfolio P containing them has risk–return

$$\overline{R}_P = R_f + \frac{\overline{R}_A - R_f}{\sigma_A}\sigma_P.$$

We want P to be efficient. This will be true if and only if A maximises the slope of this line.

Now A is a collection of individual assets; varying A is equivalent to changing the portfolio weights (i.e. investment weights) in these underlying assets. Our problem reduces to maximising the slope,

$$\theta = \frac{\overline{R}_A - R_f}{\sigma_A},$$

over the portfolio weights in A. Suppose then that x is a vector of portfolio weights, C is the covariance matrix and $\overline{R}$ is the vector of returns for the underlying assets. We have

$$\overline{R}_A = \langle x, \overline{R} \rangle, \text{ and } \sigma_A = (x^{\mathrm{T}} C x)^{\frac{1}{2}}.$$

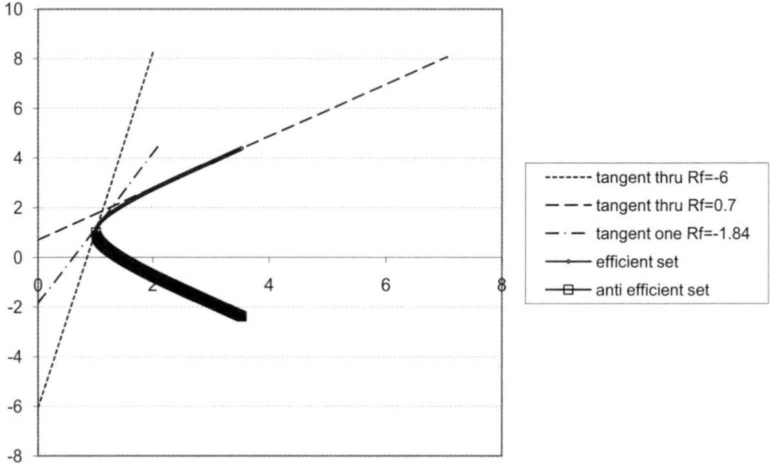

Figure 4.1 Tangent lines to the efficient hyperbola for varying risk-free rates.

Hence we must maximise

$$\theta(x) = \frac{\langle x, \overline{R}\rangle - R_f}{(x^{\mathrm{T}}Cx)^{\frac{1}{2}}} \tag{4.1}$$

on the set $\sum_{i=1}^{n} x_i = 1$.

After some calculus (which we include for those interested in the optional Section 4.5 below,) it turns out that there is an easy algorithm. To find the tangent portfolio weights we:

- let $\tilde{R}_j = \overline{R}_j - R_f$;
- solve $Cy = \tilde{R}$;
- set $x_j = \frac{y_j}{\sum_{k=1}^{n} y_k}$.

The efficient frontier is then the line through the risk-free asset and this point.

In Figure 4.1 we show some investment lines which are tangent to the efficient frontier on the hyperbola for varying risk-free rates.

4.2 Geometry of the frontier

For a given risk-free rate, we now know how to find the tangent portfolio. Since we know from Theorem 3.4 that the tangent portfolio is efficient even without the risk-free asset, we have found a portfolio that is efficient amongst the risky assets.

We will get a different efficient portfolio for each choice of risk-free rate. We can therefore find the entire efficient frontier in the risky case, simply by varying the risk-free rate.

Note that if we set

$$e = (1, 1, \ldots, 1),$$

we can rewrite the algorithm as

- set

$$\tilde{R} = \overline{R} - R_f e,$$

- put

$$y = C^{-1}\tilde{R} = C^{-1}\overline{R} - R_f C^{-1} e,$$

- let

$$x = y\langle y, e\rangle^{-1}.$$

That is, every value of y is equal to the fixed vector $C^{-1}\overline{R}$ plus a varying multiple of $C^{-1}e$. So all values of y lie on a straight line. We only need to compute $C^{-1}\overline{R}$ and $C^{-1}e$ once in order to get all values of y.

One can also see that all values of x lie on a straight line in weight space (i.e. $\mathbb{R}^n$) using a geometric argument. First, the values of y are linear combinations of $C^{-1}\overline{R}$ and $R_f C^{-1}e$ so they all lie on a straight line in $\mathbb{R}^n$. Now consider the plane which passes through this line and the origin; any line through a value of y and the origin will be contained in this plane. This immediately implies that all values of x will necessarily lie in the same plane. Furthermore, the values of x are such that $\langle x, e\rangle = 1$. This last equation defines a hyperplane in $\mathbb{R}^n$ and the intersection of a plane and a hyperplane in $\mathbb{R}^n$ will be a straight line.

We can now better understand the geometry of the efficient frontier for $n > 2$ risky assets. That is, we have that all efficient portfolios (or rather the asset weights thereof) lie on a straight line in weight space ($\mathbb{R}^n$). Noting that there is a unique straight line through two points x and w in $\mathbb{R}^n$, we also have that any point z on this straight line can be found as a linear combination of x and w, i.e

$$z = \beta x + (1 - \beta)w,$$

for some β. Hence we can find any efficient portfolio as a linear combination of x and w and we have reduced the problem to the two-asset case. It follows that the efficient frontier in mean/standard deviation space is still a hyperbola.

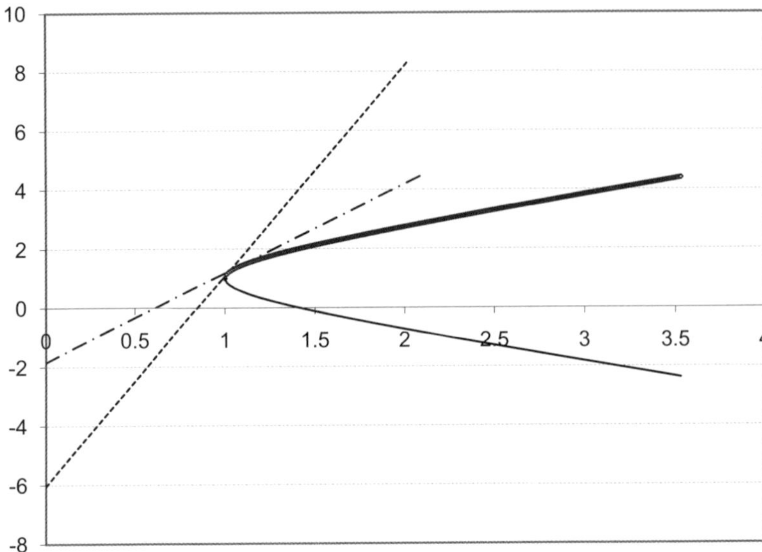

Figure 4.2 The tangent portfolio gets closer to the minimal variance portfolio as the risk-free rate becomes more negative.

4.3 The minimal variance portfolio

We turn our investigations to finding the minimal variance portfolio in the multi-asset case. From Figure 4.2, we see that as the risk-free rate gets lower and lower, the slope of the investment line gets steeper and steeper, and the tangent portfolio gets closer and closer to the tip (i.e. the point of minimal variance). It follows that we can find the weights in the minimal variance portfolio by letting the risk-free rate tend to $-\infty$.

Now, from our expressions obtained in Section 4.2 for x and y, we have:

$$\begin{aligned} x &= \frac{C^{-1}R - R_f C^{-1}e}{\langle C^{-1}R, e\rangle - R_f\langle C^{-1}e, e\rangle}, \\ &= \frac{-R_f^{-1}C^{-1}R + C^{-1}e}{-R_f^{-1}\langle C^{-1}R, e\rangle + \langle C^{-1}e, e\rangle}. \end{aligned}$$

Letting R_f tend to $-\infty$ this converges to

$$x = \frac{C^{-1}e}{\langle C^{-1}e, e\rangle}. \tag{4.2}$$

We now have, given a matrix of covariances, an easy algorithm for finding the minimal variance portfolio with n assets.

4.4 Illustrating the method

We illustrate the algorithm for finding the tangent portfolio with a numeric example and observe first that to apply the algorithm, we need to know expected returns, standard deviations and correlations, as well as the risk-free rate.

Suppose that we are given

Asset	Expected return	Standard deviation
A	15	10
B	10	6
C	20	15
R_f	3	

and pairwise correlations $\rho_{AB} = 0.4, \rho_{BC} = 0.3$, and $\rho_{AC} = 0.5$.

In order to obtain the covariance matrix, we multiply each element of the correlation matrix by the standard deviations for each of the corresponding assets:

$$\begin{pmatrix} 1\times 10\times 10 & 0.4\times 10\times 6 & 0.5\times 10\times 15 \\ 0.4\times 10\times 6 & 1\times 6\times 6 & 0.3\times 6\times 15 \\ 0.5\times 10\times 15 & 0.3\times 6\times 15 & 1\times 15\times 15 \end{pmatrix} = \begin{pmatrix} 100 & 24 & 75 \\ 24 & 36 & 27 \\ 75 & 27 & 225 \end{pmatrix}.$$

Using the quoted asset returns we have

$$\overline{R} = \begin{pmatrix} 15 \\ 10 \\ 20 \end{pmatrix},$$

and hence

$$\tilde{R} = \begin{pmatrix} 15-3 \\ 10-3 \\ 20-3 \end{pmatrix} = \begin{pmatrix} 12 \\ 7 \\ 17 \end{pmatrix}.$$

The equation to solve is therefore

$$\begin{pmatrix} 100 & 24 & 75 \\ 24 & 36 & 27 \\ 75 & 27 & 225 \end{pmatrix} \begin{pmatrix} x \\ y \\ z \end{pmatrix} = \begin{pmatrix} 12 \\ 7 \\ 17 \end{pmatrix}.$$

The solution can be found using Gaussian elimination:

$$\begin{pmatrix} x \\ y \\ z \end{pmatrix} = \begin{pmatrix} 0.05962 \\ 0.12410 \\ 0.04079 \end{pmatrix}.$$

We need the weights to add up to 1. They add up to $S = 0.2245$. So we divide by S to get

$$0.2656,\ 0.5528,\ 0.1817.$$

Recalling that

$$\mathrm{Var}\left(\sum_{i=1}^{n} X_i R_i\right) = x^{\mathrm{T}} C x,$$

we can substitute to find that the standard deviation is 6.722. Since expectation is linear, the expected return is then found as the weighted average of the underlying assets, which is 13.145.

The efficient line therefore goes through

$$(0,3) \text{ and } (6.722, 13.145).$$

The slope of the line is

$$\frac{13.145 - 3}{6.722} = 1.509,$$

and therefore the investment line has equation

$$R_P = 3 + 1.509\sigma_P.$$

It is important to realise that the algorithm actually does not guarantee that an efficient portfolio has been found. It guarantees that the portfolio found is either optimally good or optimally bad. In practice, it works when the risk-free rate is less than the expected return on the minimal variance portfolio portfolio. When it is greater, the tangent portfolio lies on the lower half of the hyperbola and the algorithm fails. The reasons for this will be clearer after reading the next section, where the algorithm is derived.

4.5 The derivation of the algorithm

This section is optional and the reader can skip it without loss of continuity. Recall that our problem is to maximise

$$\theta(x) = \frac{\langle x, \overline{R} \rangle - R_f}{(x^{\mathrm{T}} C x)^{\frac{1}{2}}} \tag{4.3}$$

on the set $\sum_{i=1}^{n} x_i = 1$, where x_i are the portfolio weights.

To find this maximum, we need to be able to do the following:

(1) find maxima of functions on $\mathbb{R}^n$, using differential calculus;
(2) find maxima with constraints via the concept of a homogeneous function;
(3) solve matrix equations.

In relation to matrix equations, we refer the reader to Appendix A for a summary of key facts about matrix algebra. We briefly review some calculus as a preliminary step. We first recall that in one-dimensional calculus, a function, f, from $\mathbb{R}$ to $\mathbb{R}$ has its maxima and minima at points where the derivative is zero. For higher dimensions, if a function f is maximal at a point

$$x = (x_1, x_2, \ldots, x_n),$$

then the one-dimensional functions

$$g_i(z) = f(x_1, \ldots, x_{i-1}, z, x_{i+1}, \ldots, x_n),$$

must have a maximum at $z = x_i$ for each i. This implies that

$$0 = g_i^{'}(x_i) = \frac{\partial f}{\partial x_i}(x) = 0,$$

for all i. If you take a derivative in any direction from a maximum, you get zero.

Example 4.1 Suppose we have a function from $\mathbb{R}^2$ to $\mathbb{R}$.

$$f(x,y) = -2x - x^2 - 3y - y^2.$$

We have

$$\frac{\partial f}{\partial x}(x,y) = -2 - 2x,$$

and

$$\frac{\partial f}{\partial y}(x,y) = -3 - 2y.$$

So at a maximum, we must have

$$x = -1, y = -3/2.$$

Thus this function has at most one maximum and it is at $(-1, -3/2)$. Note that, in general, we have to study the function to see whether the point really is a maximum. ◇

Example 4.2 Let A be a symmetric, non-singular matrix (e.g., a covariance matrix), i.e. if

$$A = (a_{ij})$$

then

$$A^{\mathrm{T}} = (a_{ji}) = A,$$

and there exists a matrix A^{-1} such that

$$AA^{-1} = I = A^{-1}A.$$

Let v be a vector and define a function f via

$$\begin{aligned} f(x) &= \langle v, x\rangle - x^{\mathrm{T}}Ax, \\ &= \sum_{j=1}^{n} v_j x_j - \sum_{i,j=1}^{n} x_i a_{ij} x_j. \end{aligned}$$

We want to find the maximum of f, and so we differentiate with respect to x_k to obtain

$$\frac{\partial f}{\partial x_k}(x) = v_k - \sum_{i=1}^{n} x_i a_{ik} - \sum_{j=1}^{n} a_{kj} x_j.$$

Since A is symmetric, this equals

$$v_k - 2\sum_{j=1}^{n} a_{kj} x_j = v_k - 2(Ax)_k.$$

At a maximum, this must be zero for all k; that is,

$$2Ax = v.$$

It follows that the only possible maximum is at the point

$$x = \frac{1}{2}A^{-1}v. \qquad \diamond$$

We next introduce the concept of rays in $\mathbb{R}^n$ and use this to solve our optimisation problem. A *ray* is a half straight-line starting at the origin. For example, all the points

$$\{(s, 2s)\}$$

with $s > 0$ form a ray in $\mathbb{R}^2$. More generally for any x, y, with at least one non-zero, the set

$$\{(xs, ys)\}$$

for $s > 0$ forms a ray in $\mathbb{R}^2$.

Every point except the origin will be on precisely one ray.

The set

$$(s, s^2) \ \forall s > 0$$

does **not** form a ray. It describes a curve not a straight line.

It will be useful to consider functions that are constant on rays. A function on $\mathbb{R}^n$ has this property, and is referred to as *homogeneous of degree zero* if and only if for all $\lambda > 0$ we have

$$f(\lambda x) = f(x).$$

Example 4.3 Define a function on $\mathbb{R}^n - \{0\}$ by

$$f(x) = \frac{x_j}{||x||},$$

where, as usual,

$$||x|| = \sqrt{\sum_{j=1}^{n} x_j^2}.$$

Then

$$f(\lambda x) = \frac{\lambda x_j}{||\lambda x||} = \frac{\lambda}{\lambda} \frac{x_j}{||x||} = f(x). \qquad \diamond$$

Given a function that is defined on a subset of $\mathbb{R}^n$, we can extend it to $\mathbb{R}^n$ by making it constant on rays.

For example, if $S = \{||x|| = 1\}$, and

$$g : S \to \mathbb{R},$$

we can define

$$f(x) = g\left(\frac{x}{||x||}\right),$$

and f is constant on rays.

If instead $S = \{\sum_{i=1}^{n} x_i = 1,\}$, similarly, let

$$h(x) = g\left(\frac{x}{\sum_{i=1}^{n} x_i}\right).$$

Then h is constant on rays and defined everywhere except the set where $\sum_{i=1}^{n} x_i = 0$.

We will want to maximise a function, f, with the constraint

$$\sum_{i=1}^{n} x_i = 1.$$

There is more than one approach to such a problem.

The one we adopt is to find a function g such that f and g agree on the set

$$E = \left\{ \sum_{i=1}^{n} x_i = 1 \right\},$$

and g is constant on rays. We then find the maxima of g instead.

Note that if g has a maximum at a point y, then by definition

$$g(z) \leq g(y)$$

for all z. But g is constant on rays so

$$g(\lambda y) = g(y), \quad \text{for all } \lambda > 0.$$

So

$$g(\lambda y) \geq g(z), \quad \text{for all } z \text{ and for all } \lambda > 0.$$

This says that if g has a maximum at the point y then every point on the entire ray through y is a maximum.

We want to maximise f on the set E. In order to do this, we first find a maximum of g and call this y. We further know that every point on the ray through y is a maximum of g. The point

$$z = \lambda y,$$

where $\lambda = \left(\sum_{i=1}^{n} y_i\right)^{-1}$, is then the point where the ray through y intersects the set E and it must still be a maximum.

Now, if $x \in E$, then $f(x) = g(x)$ since they agree on E, and $g(x) \leq g(z)$ since z is a maximum of g. So

$$f(x) \leq g(z),$$

but $f(z) = g(z)$ since f and g are equal on E so

$$f(x) \leq f(z).$$

We have shown that z is a maximum of f over the set E.

This yields a three-step approach to finding a maximum of f on the set E:

(1) find a function g which is constant on rays and agrees with f on E;
(2) find a maximum y of g;
(3) let $x = (x_1, x_2, \ldots, x_n)$ be given by

$$x_i = \frac{1}{\sum_{j=1}^{n} y_j} y_i.$$

We are now ready to address the original problem: we want to maximise

$$\theta(x) = \frac{\langle x, \overline{R} \rangle - R_f}{(x^{\mathrm{T}}Cx)^{\frac{1}{2}}} \tag{4.4}$$

on the set $\sum_{i=1}^{n} x_i = 1$. The function θ is not constant on rays. However, if we let

$$e = (1, 1, \ldots, 1),$$

then on E we have

$$\langle x, e \rangle = 1,$$

and on E, θ agrees with

$$g(x) = \frac{\langle x, \overline{R} \rangle - R_f \langle x, e \rangle}{(x^{\mathrm{T}}Cx)^{\frac{1}{2}}}. \tag{4.5}$$

Letting

$$\tilde{R} = \overline{R} - R_f e,$$

i.e. $\tilde{R}_j = \overline{R}_j - R_f$, we have

$$g(x) = \frac{\langle x, \tilde{R} \rangle}{(x^{\mathrm{T}}Cx)^{\frac{1}{2}}}. \tag{4.6}$$

We need to check that the function g is indeed homogeneous of degree zero. Let $\lambda > 0$, we have

$$\begin{aligned} \langle \lambda x, \tilde{R} \rangle &= \sum_{i=1}^{n} \lambda x_i \tilde{R}_i, \\ &= \lambda \sum_{i=1}^{n} x_i \tilde{R}_i, \\ &= \lambda \langle x, \tilde{R} \rangle. \end{aligned}$$

We also have

$$\begin{aligned} (\lambda x)^{\mathrm{T}} C (\lambda x) &= \sum_{i,j=1}^{n} (\lambda x)_i C_{ij} (\lambda x)_j, \\ &= \sum_{i,j=1}^{n} \lambda x_i C_{ij} \lambda x_j, \\ &= \lambda^2 \sum_{i,j=1}^{n} x_i C_{ij} x_j, \\ &= \lambda^2 x^{\mathrm{T}} Cx. \end{aligned}$$

So

$$((\lambda x)^{\mathrm{T}}C(\lambda x))^{1/2} = |\lambda|(x^{\mathrm{T}}Cx)^{1/2},$$

and

$$g(\lambda x) = \frac{\lambda}{|\lambda|}g(x) = g(x), \text{ for } \lambda > 0$$

and g is homogeneous of degree zero as claimed. We have completed step (1).

Now for step (2), we will use the product rule to differentiate g. We recall this for the reader who has forgotten it. For functions k and l,

$$\frac{\partial}{\partial x_i}(kl)(x) = \frac{\partial k}{\partial x_i}(x)l(x) + k(x)\frac{\partial l}{\partial x_i}(x).$$

Here we let $k(x) = \langle x, \tilde{R}\rangle$, and $l(x) = (x^t Cx)^{-1/2}$.

First, we have

$$\begin{aligned}\frac{\partial k}{\partial x_i}(x) &= \frac{\partial}{\partial x_i}\sum_{j=1}^{n} x_j \tilde{R}_j,\\ &= \tilde{R}_i.\end{aligned}$$

Second, we know $\frac{\partial}{\partial x_i}(x^{\mathrm{T}}Cx) = 2(Cx)_i$, and, using the chain rule,

$$\begin{aligned}\frac{\partial l}{\partial x_i}(x) &= -\frac{1}{2}(x^{\mathrm{T}}Cx)^{-3/2}\frac{\partial}{\partial x_i}(x^{\mathrm{T}}Cx),\\ &= -\frac{1}{2}(x^{\mathrm{T}}Cx)^{-3/2}2(Cx)_i,\\ &= -(x^{\mathrm{T}}Cx)^{-3/2}(Cx)_i.\end{aligned}$$

Putting the two terms together, we get

$$\begin{aligned}\frac{\partial g}{\partial x_j}(x) &= \frac{\tilde{R}_j}{(x^{\mathrm{T}}Cx)^{1/2}} - \frac{\langle x, \tilde{R}\rangle(Cx)_j}{(x^{\mathrm{T}}Cx)^{3/2}},\\ &= \frac{1}{(x^{\mathrm{T}}Cx)^{1/2}}\left(\tilde{R}_j - \frac{\langle x, \tilde{R}\rangle(Cx)_j}{x^{\mathrm{T}}Cx}\right).\end{aligned}$$

This will be zero if and only if

$$\tilde{R}_j = \frac{\langle x, \tilde{R}\rangle(Cx)_j}{x^{\mathrm{T}}Cx}.$$

This can be rewritten in vector form as

$$\frac{\langle x, \tilde{R}\rangle}{x^{\mathrm{T}}Cx}Cx = \tilde{R}.$$

If we let

$$\alpha(x) = \frac{\langle x, \tilde{R} \rangle}{x^{\mathrm{T}} C x}$$

then our equation becomes

$$C(\alpha(x)x) = \tilde{R}. \tag{4.7}$$

We first solve as if $\alpha(x) = 1$, and then see that it really is equal to 1. So we solve

$$Cy = \tilde{R},$$

which provided C is invertible has the unique solution

$$y = C^{-1}\tilde{R}.$$

We then have

$$\begin{aligned} \alpha(y) &= \frac{\langle C^{-1}\tilde{R}, \tilde{R} \rangle}{(C^{-1}\tilde{R})^{\mathrm{T}} C C^{-1} \tilde{R}}, \\ &= \frac{\tilde{R}^{\mathrm{T}} C^{-1} \tilde{R}}{\tilde{R}^{\mathrm{T}} C^{-1} \tilde{R}}, \\ &= 1. \end{aligned}$$

We therefore have that y is a solution of (4.7) as needed. We have completed step (2). Setting

$$x = \frac{y}{\langle y, e \rangle},$$

we have completed step (3), and we are done.

We now have a simple, straightforward algorithm for finding the tangent portfolio. That is, to find the optimal portfolio weights:

- set

$$\tilde{R}_j = R_j - R_f;$$

- solve

$$Cy = \tilde{R};$$

- set

$$x_j = \frac{y_j}{\sum_{i=1}^{n} y_i}.$$

This gives us the components of the risky portfolio so that the line through that portfolio and the risk-free asset is the efficient frontier involving the risky asset.

However, note that the derivation only guaranteed that a critical point was found: this could equally be a minimum as a maximum, so we have to be careful when applying the algorithm.

4.6 Solution via Lagrange multipliers

In this optional section, we look at how to find the efficient portfolio, with a given mean, directly. This alternative approach to the finding of efficient portfolios in the non-riskless asset case uses *Lagrange multipliers*. In particular, suppose we want to find the portfolio of minimal variance for a given expected return, μ, then we can set up a constrained optimisation problem. This problem was originally solved by Merton, [12]. Minimise

$$f(x) = \frac{1}{2} x^{\mathrm{T}} C x$$

subject to

$$g(x) = \langle x, \overline{R} \rangle = \mu \text{ and } h(x) = \langle x, e \rangle = 1.$$

The Lagrange multiplier approach involves finding a point where the gradient of f can be written as a linear multiple of the gradients of the two constraint functions. This implies that the gradient in any direction within the constrained set is zero and so the function f restricted to the constrained set has a critical point there.

So we have to find a point $x \in \mathbb{R}^n$ and two real numbers α and β such that

$$\nabla f(x) = \alpha \nabla g(x) + \beta \nabla h(x), \tag{4.8}$$

$$g(x) = \mu, \tag{4.9}$$

$$h(x) = 1. \tag{4.10}$$

Evaluating the gradients and substituting, we obtain

$$\begin{aligned} Cx &= \alpha \overline{R} + \beta e, \\ \langle x, e \rangle &= 1, \\ \langle x, \overline{R} \rangle &= \mu. \end{aligned}$$

We have $n+2$ quantities to find and $n+2$ constraints. It follows from the first equation that

$$x = \alpha C^{-1} \overline{R} + \beta C^{-1} e.$$

As in our previous derivation, we have that x is a linear multiple of $C^{-1}e$ and $C^{-1}\overline{R}$.

To find α and β we substitute the expression for x into the last two equations. Using the linearity of $\langle\cdot,\cdot\rangle$, we obtain

$$\alpha\langle C^{-1}\overline{R},e\rangle+\beta\langle C^{-1}e,e\rangle=1,$$
$$\alpha\langle C^{-1}\overline{R},\overline{R}\rangle+\beta\langle C^{-1}e,\overline{R}\rangle=\mu.$$

We now have a 2×2 system,

$$\begin{pmatrix}\langle C^{-1}\overline{R},e\rangle & \langle C^{-1}e,e\rangle\\ \langle C^{-1}\overline{R},\overline{R}\rangle & \langle C^{-1}e,\overline{R}\rangle\end{pmatrix}\begin{pmatrix}\alpha\\ \beta\end{pmatrix}=\begin{pmatrix}1\\ \mu\end{pmatrix}.$$

This is straightforward to solve for α and β and we are done. It can be shown that if C is positive definite then C^{-1} is also positive definite and it then follows that this system always has a unique solution.

4.7 Review

By the end of this chapter, the reader should be able to do the following questions and tasks.

1 How do you find the tangent portfolio in the multi-asset case?
2 How do you find the minimal variance portfolio in the multi-asset case?
3 Sketch the efficient frontier in the multi-asset case in return/standard deviation space.
4 Describe the geometry of the efficient frontier in weight space with and without a risk-free asset.
5 Solve problems that involve finding a portfolio prescribed expected return or standard deviation in the multi-asset case.
6 What data is required to compute the tangent portfolio?
7 Give the algorithm for finding the tangent portfolio.
8 Give the algorithm for finding the minimal variance portfolio.
9 What shape does the efficient frontier take if there are $n>2$ risky assets and no risk-free assets in weight space and in expected return/standard deviation space?
10 What shape does the efficient frontier take if there are $n>2$ risky assets and a risk-free asset in weight space and in expected return/standard deviation space?

4.8 Problems

Question 4.1 Give an example of a symmetric matrix that contains no zeros and is not invertible.

Question 4.2 Two assets have mean returns 11 and 12, standard deviations of 10 and 11, and correlation 0. Find the tangent portfolio and minimal variance portfolio (excluding the risk-free asset) if the risk-free rate is 1. Find also the equation of the investment line.

Question 4.3 Two assets have mean returns 11 and 12, standard deviations of 6 and 7, and correlation 0. Find the tangent portfolio and minimal variance portfolio (excluding the risk-free asset) if the risk-free rate is 5. Find also the equation of the investment line.

Question 4.4 Suppose assets A, B, C have expected returns 11, 14 and 17. Suppose their covariance matrix of returns is

$$\begin{pmatrix} 1 & 1 & 1 \\ 1 & 4 & 1 \\ 1 & 1 & 2 \end{pmatrix}.$$

Find the composition of the minimum variance portfolio involving A, B and C. What are its expected return and variance? If the risk free rate is 5, find the composition of the tangent portfolio, and compute its expected return and standard deviation.

Question 4.5 Suppose assets A, B, C have expected returns 12, 16 and 8. Suppose their covariance matrix of returns is

$$\begin{pmatrix} 2 & 1 & 0 \\ 1 & 3 & 1 \\ 0 & 1 & 2 \end{pmatrix}.$$

Find the composition of the minimum variance portfolio involving A, B and C. What is its expected return and variance?

Question 4.6 Let e be a vector of 1s. Show that if $\overline{R}$ is the vector of expected returns, R_f is the risk-free rate, $x = \overline{R} - R_f e$, C is the covariance matrix of returns and $y = C^{-1}x$, then the tangent portfolio, v, has variance

$$\frac{1}{\langle y, e \rangle} x^{\mathrm{T}} v.$$

Question 4.7 What is the limit as x goes to infinity of

$$\frac{\alpha + \beta x + \gamma x^2}{\delta + \varepsilon x + \nu x^2}?$$

Question 4.8 Suppose risky assets have mean returns 11, 12, 13, 14, 15, 16. There exists a risk-free asset with return 5. Investors A and B are mean–variance investors. Investor A holds the fractions 0.1, 0.11, 0.12, 0.13, 0.14, 0.15, of his portfolio in each of the risky assets respectively. Investor B holds 0.25 of the first asset. How much does she hold of each of the other assets?

Question 4.9 An investor can invest in any of 250 stocks. He can place money on deposit at a rate of 3% and cannot borrow. If he is a mean–variance investor, describe the geometry of the set of investments he would consider in:

- the space of portfolio weights;
- expected return/standard deviation space.

Question 4.10 An investor can invest in n assets, X_j, with returns $R_1,\dots,R_n$. A portfolio with weights $\theta_1,\theta_2,\dots,\theta_n$, is known to be mean–variance efficient. The variance of R_n is not zero. A second investor cannot invest in X_n. Will the portfolio with weightings

$$\phi_j = \theta_j / \sum_{i<n} \theta_i$$

be mean–variance efficient for this investor, in general? Justify your answer.

Question 4.11 An investor can invest in n assets, X_j, with returns $R_1,\dots,R_n$. A portfolio with weights $\theta_1,\theta_2,\dots,\theta_n$, is known to be mean–variance efficient. The variance of R_n is zero. A second investor cannot invest in X_n. Will the portfolio with weightings

$$\phi_j = \theta_j / \sum_{i<n} \theta_i$$

be efficient for this investor, in general? Justify your answer.

Question 4.12 Investments A, B, C and D have covariance matrix of returns

$$\begin{pmatrix} 4 & 3 & 3 & 0 \\ 3 & 4 & 3 & 0 \\ 3 & 3 & 3 & 0 \\ 0 & 0 & 0 & 0 \end{pmatrix},$$

and have expected rates of return 6, 5, 4, and 2 respectively. Find the composition and standard deviation of an efficient portfolio with expected rate of return 6.75.

Question 4.13 Investments A, B, C, D and E have covariance matrix of returns equal to

$$\begin{pmatrix} 5 & 2 & 2 & 1 & 3 \\ 2 & 4 & 2 & 3 & 1 \\ 2 & 2 & 4 & 2 & 2 \\ 1 & 3 & 2 & 4 & 2 \\ 3 & 1 & 2 & 2 & 4 \end{pmatrix},$$

and have expected rates of return 7, 7, 13, 13, and 13, respectively. The risk-free asset has rate of return 1. Investors X, Y, and Z are mean–variance investors who can hold unlimited long and short positions in these assets. We know that X holds a portfolio of -1 units of B, 1 unit of C, 1.5 units of D, 0.5 units of E and -1 units of the risk-free asset. What portfolio should be held:

1. by Y to achieve an expected rate of return equal to 8.5;
2. by Z to achieve a variance of 30?

5
Single-factor models

5.1 Introduction

We have seen that given the covariances and expected returns of a collection of securities, then we can find the tangent portfolio and hence the efficient frontier – that is, the set of portfolios which are either optimal in mean return for the level of variance, or in variance for the level of mean return.

However, one problem with this approach in practice is the large amount of data required. To see this, suppose we have N assets. We need the covariance matrix of returns and the mean values of returns for each of these assets. This requires

$$N + N(N+1)/2$$

numbers and the number of entries needed grows with the square of the number of assets.

A typical market will have at least 150 securities and to achieve maximal diversification, we should be considering as many securities and therefore markets as possible. If we were to consider 500 securities, then we would need 500 pieces of information for the estimation of returns, and $500^2 = 250,000$ for the covariance matrix.

If this is not daunting enough, we are soon confronted with another problem, namely how to obtain these numbers. One standard method is historical analysis where estimates are made from historical time series. This however is fraught with its own issues, not least of which is that for each of the $N + N(N+1)/2$ quantities, a large number of data points are required to make a sensible estimate. Furthermore, since we are generally interested in 1-year time horizons, we find ourselves needing to go back a long way in time. However, we do not have to go back all that far to find that both our companies (if they existed at all) and the markets in which they operated were significantly different from today. The validity of this historical data is therefore highly questionable.

One solution may be to use shorter time horizons and to scale accordingly. Yet evidence exists that short term behaviour is qualitatively different and this makes this approach hard to justify. Another possibility may be to use professional analysts' estimates for each of the quantities. To make these estimates, company analysts examine specific company and sector information and use this to try to predict profitability and other key performance measures. However, even without the obvious issues of reliability – analysts' forecasts may and do differ from each other and from that which is actually achieved – the number of analysts required to estimate $500,000$ data points (and the cost thereof) would certainly outweigh the benefits.

For these reasons, we need a simpler model whilst retaining the large number of assets, since more assets leads to diversification and hence less risk for the level of return. A standard approach is to us a *single-factor model* in which all asset correlation arises from their common exposure to a single index, in this case that of the overall market. We note that here one factor means one **common** factor which is quite different from having **one factor in total.**

5.2 Mathematical formulation of the single-factor model

Firstly, suppose that R_m is the return on the market index that is a synthetic portfolio in which all market assets are represented in the proportions they occur. For each asset R_i, we set

$$R_i = \alpha_i + \beta_i R_m + e_i,$$

where α_i and β_i, are constants and the random variables e_i have mean zero and are uncorrelated with the market.

This does not say anything until we make the crucial additional assumption:

$$\mathbb{E}(e_i e_j) = 0, \text{ for } i \neq j;$$

that is to say, the variables e_i are not correlated with each other. We note that since their means are zero, this constraint is equivalent to zero correlation.

From these assumptions, we are able to deduce the following key theorem.

Theorem 5.1 *In a single-factor model:*

(1) $\mathbb{E}(R_i) = \beta_i \mathbb{E}(R_m) + \alpha_i$;
(2) $\mathrm{Var}(R_i) = \beta_i^2 \sigma_m^2 + \sigma_{e_i}^2$;
(3) *the covariance between security i and security j is* $\beta_i \beta_j \sigma_m^2$, *for* $i \neq j$.

The first result follows directly from linearity of expectations and our assumption that $E(e_i) = 0$. That is, we have

$$\begin{aligned}\mathbb{E}(R_i) =&\mathbb{E}(\alpha_i + \beta_i R_m + e_i),\\ =&\alpha_i + \beta_i \mathbb{E}(R_m) + \mathbb{E}(e_i),\\ =&\alpha_i + \beta_i \mathbb{E}(R_m).\end{aligned}$$

The second draws on the assumption that each e_i is uncorrelated with the market, and for uncorrelated random variables X and Y, we have

$$\mathrm{Var}(X+Y) = \mathrm{Var}(X) + \mathrm{Var}(Y).$$

We therefore have

$$\begin{aligned}\mathrm{Var}(\alpha_i + \beta_i R_m + e_i) =&\mathrm{Var}(\beta_i R_m) + \mathrm{Var}(e_i),\\ =&\beta_i^2 \,\mathrm{Var}(R_m) + \mathrm{Var}(e_i).\end{aligned}$$

Note also that since α_i is constant, it has no effect on variance.

Finally, to derive the expression for $\mathrm{Cov}(R_i, R_j)$, we have from our definitions

$$\begin{aligned}\mathrm{Cov}(R_i R_j) =&\mathbb{E}((R_i - \overline{R}_i)(R_j - \overline{R}_j)),\\ =&\mathbb{E}((\beta_i(R_m - \overline{R}_m) + e_i)(\beta_j(R_m - \overline{R}_m) + e_j)).\end{aligned}$$

Expanding and applying linearity of expectation, this reduces to the sum of four terms. These are

$$\begin{aligned}&\beta_i\beta_j\mathbb{E}((R_m - \overline{R}_m)^2),\\ &\beta_i\mathbb{E}((R_m - \overline{R}_m)e_j),\\ &\beta_j\mathbb{E}((R_m - \overline{R}_m)e_i),\\ &\mathbb{E}(e_i e_j).\end{aligned}$$

However, for $i \neq j$, the last three terms are equal to zero from our assumption that the terms e_i are uncorrelated both with the market and each other.

5.3 Data requirements for the single-factor model

Assuming a single-factor model of the above form, our data requirements reduce significantly. That is to say, the problem reduces to obtaining estimates for the quantities:

$$\alpha_i, i = 1, 2, \ldots, n,$$

$$\beta_i, i = 1,2,\ldots,n,$$

$$\mathbb{E}(R_m),$$

$$\text{Var}(R_m);$$

and the standard deviations of the residuals:

$$\sigma_{e_i}, \quad i = 1,2,\ldots,n.$$

Accordingly, for N securities, this means that we need to estimate $3N+2$ quantities. Importantly, this number will grow linearly with the number of assets, rather than quadratically. Crucially, by constraining the covariance structure with the zero correlation assumption for the terms e_i, we have vastly reduced the amount of data required.

5.4 Understanding beta

Returning to our expression for $\text{Var}(R_i)$, we see that the single-factor model divides the risk for the security into two pieces.

- The first part arises from exposure to the market and is called *systematic risk.*
- The remaining part is called *specific risk.* It is the part unique to the security and it can be diversified away.
- Specific risk is also referred to as alpha risk, diversifiable risk, unsystematic risk, and residual risk.

Clearly,

$$R_m = 0 + 1.R_m,$$

and hence the market portfolio has

$$\begin{aligned}\alpha &= 0,\\ \beta &= 1.\end{aligned}$$

The level of beta indicates the risk level compared to the market. For example, suppose that a stock has a beta of 3 and an alpha of 0. Suppose further that the market goes up by 1%. What can we say about the stock's return? Our best estimate without further knowledge is that it will have gone up by 3% on average. However, in fact, any value is possible because of residual risk.

We will show shortly that by holding a sufficiently large number of securities we can effectively eliminate our exposure to residual risk. It follows that since we can diversify this risk we should not expect to obtain any risk premium for holding it. Thus it is only the β that leads to a risk premium: we only receive reward for our exposure to the overall level of the market.

We start by considering a portfolio of N assets, with weights X_i such that $\sum_{i=1}^{n} X_i = 1$. The portfolio risk may be computed as follows:

$$\begin{aligned}
&\mathrm{Var}\left(\sum_{i=1}^{n} X_i R_i\right) \\
&= \mathbb{E}\left(\left(\sum_{i=1}^{n} X_i (R_i - \overline{R}_i)\right)^2\right), \\
&= \mathbb{E}\left(\left(\sum_{i=1}^{n} X_i \beta_i (R_m - \overline{R}_m) + \sum_{i=1}^{n} X_i e_i\right)^2\right), \\
&= \mathbb{E}\left(\left(\sum_{i=1}^{n} \beta_i X_i\right)^2 (R_m - \overline{R}_m)^2\right) + \sum_{i=1}^{n} X_i^2 \mathbb{E}(e_i^2), \\
&= \left(\sum_{i=1}^{n} X_i \beta_i\right)^2 \sigma_m^2 + \sum_{i=1}^{n} X_i^2 \sigma_{e_i}^2.
\end{aligned}$$

The beta of the portfolio will be

$$\sum_{i=1}^{n} X_i \beta_i,$$

which is the weighted average of the betas of the constituent stocks.

Suppose then that we divide our portfolio into large number, N, of assets with equal amounts of money in each. Substituting $X_i = \frac{1}{N}$ into the above expression, we have that the portfolio variance is

$$\sigma_P^2 = \beta_P^2 \sigma_m^2 + \frac{1}{N} \overline{\sigma_{e_i}^2}, \tag{5.1}$$

where the term β_P is the average of the asset betas, and $\overline{\sigma_{e_i}^2}$ is the average of the idiosyncratic variances.

Provided that the individual risks $\sigma_{e_i}^2$ are bounded, the second term $\frac{1}{N}\overline{\sigma_{e_i}^2}$ will go to zero as N gets bigger. (That is, for any arbitrarily small ε there will exist a number k for which this term is less than ε for all $N > k$.)

It follows that by using a large portfolio of many different stocks, one can make the diversifiable risk disappear. In other words, the residual or specific risk for each of the stocks is diversified away.

However, some risk does remain and that is the beta or undiversifiable risk. We would therefore expect to receive risk premia for taking beta risk but not diversifiable risk.

We further note from equation (5.1) that the remaining residual risk is proportional to $1/N$. Hence even with as few as 10 stocks the residual risk is reduced to 10% of the average for the individual stocks. For a portfolio of 100 stocks, the residual risk is only 1% of the average.

5.5 Techniques for parameter estimation

We have developed a simplified model of asset returns. We have fewer parameters than before but we still need to fit those parameters to the market. That is, we still need a methodology for finding the α and β coefficients. We approach this by first examining a more general problem.

Given random variables X and Y, how do we find α and β such that

$$e = Y - (\alpha + \beta X)$$

has zero mean and is uncorrelated with X?

The residual, e, will have zero expectation and zero covariance with X if and only if

$$\mathbb{E}(e) = \mathbb{E}(Y) - \alpha - \beta\mathbb{E}(X) = 0, \tag{5.2}$$

$$\mathbb{E}(eX) = \mathbb{E}(YX) - \alpha\mathbb{E}(X) - \beta\mathbb{E}(X^2) = 0, \tag{5.3}$$

where we are using the fact that

$$\text{Cov}(e, X) = \mathbb{E}(eX)$$

if and only if $\mathbb{E}(e) = 0$.

We can rewrite the above equations as a linear system as follows:

$$\begin{pmatrix} 1 & \mathbb{E}(X) \\ \mathbb{E}(X) & \mathbb{E}(X^2) \end{pmatrix} \begin{pmatrix} \alpha \\ \beta \end{pmatrix} = \begin{pmatrix} \mathbb{E}(Y) \\ \mathbb{E}(XY) \end{pmatrix}.$$

It then remains to solve the system by inverting the matrix. Since it is a 2×2 matrix, the inverse is found via the standard formula:

$$\frac{1}{\mathbb{E}(X^2) - \mathbb{E}(X)^2} \begin{pmatrix} \mathbb{E}(X^2) & -\mathbb{E}(X) \\ -\mathbb{E}(X) & 1 \end{pmatrix} = \frac{1}{\text{Var}(X)} \begin{pmatrix} \mathbb{E}(X^2) & -\mathbb{E}(X) \\ -\mathbb{E}(X) & 1 \end{pmatrix}.$$

Left multiplying the inverse on both sides of the matrix equation, we obtain

$$\beta = \frac{1}{\text{Var}(X)} \left(\mathbb{E}(XY) - \mathbb{E}(X)\mathbb{E}(Y)\right);$$

that is,

$$\beta = \frac{\mathrm{Cov}(X,Y)}{\mathrm{Var}(X)}.$$

For our application, $R_S = Y$ will be the return of the stock, and $R_m = X$ will be the market return.

Of course, we will not actually know the distributions of X and Y. All we will have available to us will be historical estimates of the means, variances, and covariances. We will use the above result to formulate a procedure for estimating beta using observed returns for the stock and the market.

We define our procedure as follows:

- let $R_{S,i}$ and $R_{m,i}$ denote the returns for stock S and the market for the ith period;
- compute the average returns of the stock, $\overline{R}_S$, and the market, $\overline{R}_m$;
- let

$$\mathrm{Cov}(R_S, R_m) = \frac{1}{N}\sum_{i=1}^{N}(R_{S,i} - \overline{R}_S)(R_{m,i} - \overline{R}_m);$$

- let $\mathrm{Var}(R_m) = \frac{1}{N}\sum_{j=1}^{N}(R_{m,j} - \overline{R}_m)^2$;
- let

$$\beta = \frac{\mathrm{Cov}(R_S, R_m)}{\mathrm{Var}(R_m)}.$$

The easiest way to compute α is first to compute β, and then, using

$$\mathbb{E}(R_S - (\alpha + \beta R_m)) = 0,$$

we have

$$\alpha = \mathbb{E}(R_S - \beta R_m) = \frac{1}{N}\sum_{j=1}^{N}\left(R_{S,j} - \beta R_{m,j}\right).$$

An alternative (and equivalent) way of finding α and β is to view e as the residual after least-squares regression of Y upon X. In other words, we want to find α and β which minimise the expression

$$\mathbb{E}((Y - (\alpha + \beta X))^2).$$

This expresses the idea that α, β are chosen to make $\alpha + \beta X$ the best possible estimate of Y; that is, the error is minimised. This expression is quadratic in α and β and its minimum will occur at a point where its derivatives with respect to α and β are zero.

α	1%,
β	2,
R_m standard deviation	2.5%,
e_i standard deviation	2%,
R_m mean	3%.

Table 5.1 *Parameters for a single-factor model.*

The derivative with respect to α is

$$2\mathbb{E}(Y-(\alpha+\beta X))=2\mathbb{E}(e),$$

and the derivative with respect to β is

$$2\mathbb{E}((Y-\alpha+\beta X)X)=2\mathbb{E}(\langle e,X\rangle).$$

When $\mathbb{E}(e)=0$, $\mathrm{Cov}(e,X)=\mathbb{E}(\langle e,x\rangle)$, so both derivatives are zero precisely when $\mathbb{E}(e)=0$ and $\mathrm{Cov}(e,X)=0$. The reader will recognise that this is just the same as the linear system in equation (5.2) and we have arrived at the same point from a different approach.

Linear regression almost always works in the sense that it will always find a straight line through a cloud of points that minimises the least-squares error. The residual generated will be uncorrelated with the market and will have zero expectation. However, the residuals may well have high correlation with each other.

A key assumption of the single-factor model is that the residuals are uncorrelated with each other. It follows that when they are highly correlated the model is a poor fit to the data, and one should consider carefully whether it should be used.

5.6 Assessing estimates

It is important to realise that the estimates by their very nature may contain a great deal of noise. Even if the model were wholly correct, our computation of the values of α and β based on historically observed data would still contain errors.

We may examine our estimates first by assuming we have a true model with parameters as in Table 5.1. Suppose we then use this data to generate a time series and then measure the implied α and β. What happens?

An example of such a series so-generated is given in Table 5.2. If we compute mean return on the market we get 3.05%, and for the stock we get 7.13%.

market return	stock return
6.430%	14.48%
3.998%	8.89%
2.350%	5.57%
0.506%	2.43%
2.431%	5.44%
8.445%	14.79%
0.624%	–1.49%
2.484%	11.01%
0.151%	3.08%

Table 5.2 *A time series from the single-factor model of Table 5.1.*

market return	stock return
3.38%	7.35%
0.95%	1.76%
–0.70%	–1.56%
–2.54%	–4.70%
–0.62%	–1.69%
5.40%	7.66%
–2.42%	–8.62%
–0.56%	3.88%
–2.90%	–4.05%

Table 5.3 *A time series from the single-factor model of Table 5.1 with the means subtracted.*

Working with the same synthetic data set we may then subtract the means to obtain the results as shown in Table 5.3.

The covariance is obtained by taking the product of the values in the first and second columns and dividing by $N = 9$ (being the number of observation periods). The result is 0.1239%. The variance of the market is obtained by taking the sum of the squares and dividing by $N = 9$. The result is 0.0704%.

The beta is then the ratio of the covariance to the variance and we find

$$\beta = 1.77.$$

Substituting, we then have

$$\alpha = \mathbb{E}(S) - \beta\mathbb{E}(R_m) = 1.73\%.$$

Clearly, our estimates of α and β are not very good even in the highly artificial situation in which we assume that the model itself is correct. The reader may well note that this is just one set of numbers. For a more thorough examination, we would do this many times. An example using 12 generated time series is given in Table 5.4. (The α numbers are percentages.)

alpha	beta
1.17	1.93
1.77	1.75
1.51	1.78
1.91	1.72
1.57	1.78
1.76	1.81
2.37	1.56
2.33	1.51
0.06	2.34
1.12	2.03
–1.08	2.65
–6.50	4.51

Table 5.4 *Values of α and β implied by taking time series using parameters from Table 5.1*

alpha	beta
1.10	1.86
0.85	1.97
1.68	1.92
1.20	2.00
0.76	2.06
0.78	1.99
1.24	1.90
1.22	1.84
1.39	1.91
0.60	2.02

Table 5.5 *Values of α and β implied by taking time series using parameters from Table 5.1 with monthly data for 10 years.*

Much variation is evident. Some of the values of β are quite good but some are terrible compared to the assumed value of 2. Importantly, when using real data, it is impossible to know which are the better values.

How can we get less noisy estimates? More data points would seem like the obvious solution. We could perhaps use monthly returns instead of yearly returns to estimate α and β. This will give us less noise but the estimate it provides will be of month-on-month rather than year-on-year correlation. These will be similar but there is evidence that they are different.

We give examples of such computations in Table 5.5. We see that they are much more stable.

5.7 Portfolio betas

The size of the idiosyncratic (or residual) risk, i.e. that arising from the e_i components is an important factor in determining the accuracy of the resulting beta estimates. In particular, if we take a portfolio that has had much of the idiosyncratic risk diversified away, the beta estimate will be much better. The more similar the portfolio is to the market, the less residual risk there will be.

We can compute

$$\begin{aligned}\hat{\beta} = \frac{\operatorname{Cov}(X,Y)}{\operatorname{Var}(X)} =& \frac{\operatorname{Cov}(X, \alpha+\beta X+e)}{\operatorname{Var}(X)}, \\ =&\beta\frac{\operatorname{Var}(X)}{\operatorname{Var}(X)} + \frac{\operatorname{Cov}(e,X)}{\operatorname{Var}(X)}, \\ =&\beta + \frac{\operatorname{Cov}(e,X)}{\operatorname{Var}(X)}.\end{aligned}$$

These expressions will hold not just for the true values, but also for sample estimates. So $\hat{\beta}$, the sample beta, is the true beta plus the sample $\operatorname{Cov}(e,X)$. Thus when e has small variance, the sample value of $\operatorname{Cov}(e,X)$ will be close to its true value of zero, and the error will be small.

5.8 Blume's technique

If we observe a large sample of stocks and compute the beta for each, we will get a distribution of betas. Each of these betas will have an error. If the observed beta is very high then the high value is as likely to come from upwards error as from the true value being high. Similarly, for very low betas, it is as likely to come from a large downwards error as from the true value being low. This suggests that betas should have a tighter distribution than typically observed. In other words, they should be closer to one. This is an example of the statistical phenomenon of "regression to the mean." We illustrate this in Figures 5.1 and 5.2 and the following example.

Example 5.2 Suppose we believe that the betas have a normal distribution with mean 1 and standard deviation 0.1. Suppose we believe that our estimate error, E, is normal with standard deviation 0.3 and independent of β.

We observe for some stock S_i that its beta is

$$\hat{\beta}_i = 1.8.$$

Intuitively, we would expect this to be an overestimate, since the probability of an 8 standard deviation draw is vanishingly small for a normal distribution. ◇

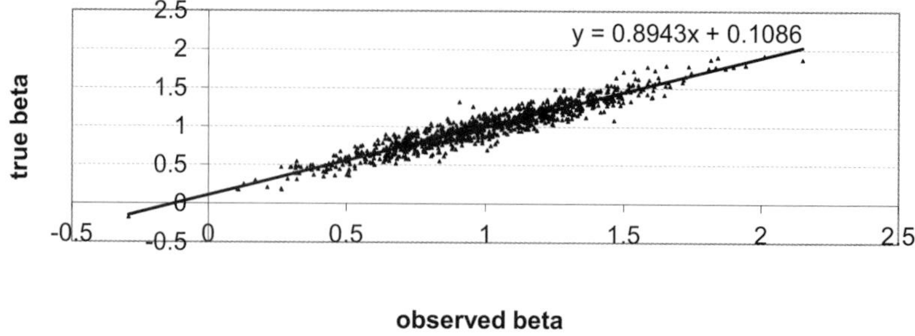

Figure 5.1 Regression of true beta against observed betas when true betas have a standard deviation of 0.3 and errors have a standard deviation of 0.1.

Regression line of true beta against observed one: high error

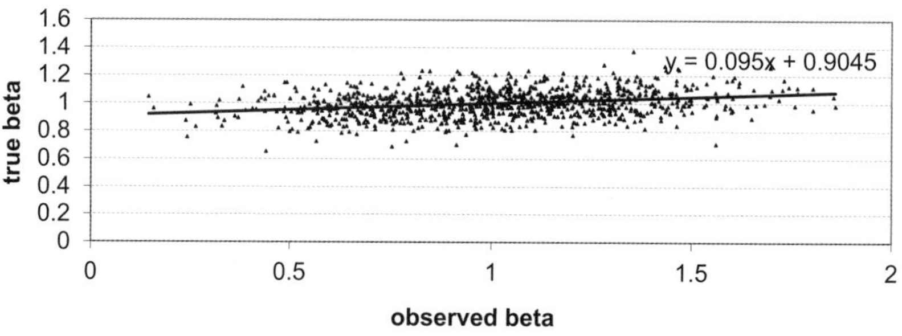

Figure 5.2 Regression of true beta against observed betas when true betas have a standard deviation of 0.1 and errors have a standard deviation of 0.3.

Suppose now that we have N stocks and that their true betas can be represented by independent draws from a probability distribution. We can represent a draw by a random variable β_X. Suppose also that our observation has an error, ε, which is independent of the beta.

Our collection of observed betas is then a collection of observations of the form $\beta_X + \varepsilon$. Since they are independent we have

$$\operatorname{Var}(\beta_X + \varepsilon) = \operatorname{Var}(\beta_X) + \operatorname{Var}(\varepsilon) > \operatorname{Var}(\beta_X).$$

So the observed distribution has larger variance than the true one.

Our objective is to find a good prediction of β for the next period. It will therefore help to remove the extra variance.

Period	Preceding period	Regression
July 33 – June 40	July 26 – June 33	$\beta_2 = 0.320 + 0.714\beta_1$
July 40 – June 47	July 33 – June 40	$\beta_2 = 0.265 + 0.750\beta_1$
July 47 – June 54	July 40 – June 47	$\beta_2 = 0.526 + 0.489\beta_1$
July 54 – June 61	July 47 – June 54	$\beta_2 = 0.343 + 0.677\beta_1$
July 61 – June 68	July 54 – June 61	$\beta_2 = 0.399 + 0.546\beta_1$

Table 5.6 *Regression of observed value of beta, β_2, against observed value of beta, β_1, in preceding period*

If the distribution of observed betas is too wide then:

- observed betas that are high should be overestimates;
- observed betas that are low should be underestimates.

Suppose then that we measure betas of stocks in one period, $\beta_{1,i}$, and then remeasure them for the same stocks in the next period, $\beta_{2,i}$, with $i = 1, 2, \ldots, N$.

If we are correct then high betas should tend to go down, and low betas should tend to go up. We may use regression to test this. If we regress to get

$$\beta_{2,i} \sim a\beta_{1,i} + b,$$

where i ranges over all the stocks, and a and b are the same for all stocks, then we should get $a < 1$. Blume, [1]. tested this hypothesis. He measured the value of β for a range of stocks in a number of periods. Each period was seven years long. This allowed enough data for each period to obtain a reasonable but not perfect estimate of the beta for that period. We thus have one beta for each stock for each period. He regressed the observed value in each period against the observed value in the preceding period. He found that the straight line had slope less than 1, which indeed suggested that the initial estimates were too far apart. Blume's results are shown in Table 5.6.

For example, if the observed value of beta in 54 to 61 was 1.4, then our best estimate from 61 to 68 is

$$1.16 = 0.399 + 0.546 \times 1.4.$$

Now suppose we are working with a fund that has a five-year time horizon. We want to make the best estimate we can of the beta of each stock for the next five years. Since we are unable to regress against the future, we therefore make the modelling assumption that the ten-year period starting five years ago is the same as the ten-year period starting ten years ago. In practice this means: we regress betas from 0 to 5 years ago, β_i^{last}, against ones from 5 to 10 years,

$\beta_i^{\text{previous}}$, to get a, b so that

$$\beta_i^{\text{last}} = a\beta_i^{\text{previous}} + b,$$

is as well-fitted as possible.

Once we have a and b, we then use β for the last five years for each stock, β_i^{last}, and we put

$$\beta_i^{\text{next}} = a\beta_i^{\text{last}} + b,$$

to estimate for the next five years.

Blume's method is very *ad hoc* in nature. It does not have a firm mathematical foundation, but has value as a generally plausible idea. It does not take into account that for stocks with low idiosyncratic risk, the beta will be more accurate, which would suggest less scaling is required for such stocks. Vasicek [28] has suggested a Bayesian technique that corrects this deficiency, but as with all Bayesian techniques it requires the user to have a *prior* distribution: that is, a view on what distribution of βs is reasonable. We will not go into the details here.

Another issue with Blume's method is that it does not ensure that a portfolio with $\beta = 1$ and no idiosyncratic risk, i.e. the market!, is mapped to a portfolio with $\beta = 1$.

We illustrate Blume's method with an example.

Example 5.3 Suppose we got the same numbers as Blume's last regression and found

$$\beta_{05\text{–}09,i} = 0.399 + 0.546\beta_{00\text{–}04,i}.$$

We can use this to predict $\beta_{10\text{–}14,i}$. So if the betas of four portfolios in 05–09 are 0.9, 1.05, 1.5 and 2 then we estimate their values for 10–14 via

$$\begin{aligned}
0.9 &\mapsto 0.399 + 0.546 \times 0.9 = 0.8904,\\
1.05 &\mapsto 0.399 + 0.546 \times 1.05 = 0.9723,\\
1.5 &\mapsto 0.399 + 0.546 \times 1.5 = 1.218,\\
2 &\mapsto 0.399 + 0.546 \times 2 = 1.491.
\end{aligned}$$

Note that 0.9 has actually moved out slightly further from 1 since we have not pinned 1 to map to 1. In fact, it maps to 0.945. ◇

5.9 Fundamental analysis

Another approach for estimating β is to consider where β comes from. What makes one company riskier than another? One view is that it comes from the fundamental characteristics of the firm. If we define beta in terms of these, we get a measure that is much more reactive. If the company risks up its strategy through leverage, for example by issuing a large amount of debt, it will immediately be reflected in the beta. By contrast, on applying the historical method it would take years for the increased risk to filter through to the measured values.

In order to build such a fundamental model for forward-looking beta, we would need to first consider what sort of quantities could be regarded as important? Some examples might be:

- dividend to earnings ratio;
- asset growth, i.e., percentage increase;
- leverage, e.g. debt to equity ratio;
- earnings variability;
- accounting beta: historical covariance of earnings against market returns;

many others are conceivable.

We would also need to work out how to use these fundamental characteristics in practice. To do this, we would need to understand just how these characteristics affect the beta. Accordingly, an approach could be outlined as follows:

- measure betas over some past period e.g., 5 years;
- carry out a linear regression of the betas against the fundamental characteristics;
- use the regressed equation to predict the future beta for each firm.

5.10 Review

By the end of this chapter, the reader should be able to do the following questions and tasks.

1. What are the data problems with mean–variance analysis?
2. What is a single-factor model?
   - Derive expressions for expected return, variance and covariance.
   - Define specific risk, systematic risk and diversifiable risk.
   - Derive expressions for the beta and variance of a portfolio.
   - What happens to the variance of a large portfolio as the number of assets goes to infinity?
3. Show how to find the β of a stock in a single-factor model, given variances and covariances of returns.
4. How do we find the α of a stock from market data in a single-factor model?

5. How can we improve the stability of α and β estimates when using a single-factor model?
6. How could you assess if a single-factor model is suitable for fitting a covariance matrix?
7. If we linearly regress the betas in one period against a previous period, what properties would we expect of the coefficients found?
8. Describe Blume's technique and discuss briefly why it is plausible.
9. What is the advantage of using fundamental analysis to estimate betas?

5.11 Problems

Question 5.1 Suppose a single-factor model is an accurate model of security returns, and that in the model every stock has the same residual risk term with standard deviation σ_{e_i}. What standard deviation of residual risk would you expect in portfolios of 10, 20 and 50 equally weighted stocks?

Question 5.2 Suppose a single-factor model is an accurate model of security returns, and that in the model every stock has the average standard deviation of residual risk is 20. How many stocks would be needed to reduce the standard deviation of residual risk to 10, 1, and 0.1?

Question 5.3 We model 3 stocks using a single-factor model, they have betas of 3, 2, and 0.5, and their alphas are 1, 2, and 3. Their residual risks have variances 10, 20, and 30. The market has expected return 10 and variance 20. Compute the stocks' expected returns and covariance matrix of returns.

Question 5.4 Suppose investments R_j have the distribution

$$\alpha_i + \beta_j R_m + f_j,$$

where R_m is the return on the market, and the terms f_j have the properties

$$\mathbb{E}(f_i f_j) = \delta \text{ for } i \neq j,$$

and $\mathbb{E}(f_i) = 0$, and $\mathbb{E}(f_i^2) = \varepsilon$. They are also independent of R_m. Find expressions for

- the expected return of asset i,
- the variance of asset i,
- the covariance of asset i and asset j.

Question 5.5 The risk-free rate is 3. The market has expected return 10 and variance of returns 4. Three assets are modelled by a single-factor model with alphas 1, 2, and 2; betas 0.5, 1, and 2; and variance of residual risk 1, 1, and 1. Find the tangent portfolio for a mean–variance investor. What are its expected return and standard deviation of returns?

Question 5.6 The risk-free rate is 2. The market has expected return 12 and variance of returns 12. Three assets are modelled by a single-factor model with alphas $10, 6$ and 4, betas $2, 1$ and 0.5 and variance of residual risk $1, 1$ and 1. Find the optimal portfolio for a mean-variance investor who wishes a variance of 10. What are its expected return and standard deviation of returns?

Question 5.7 Suppose the market has expected return 10, and standard deviation 5. Two stocks have expected returns 15 and 20, standard deviations 10 and 15. Their correlations with the market are 0.1 and 0.2. Assuming a single-factor model, find their alphas and betas.

Question 5.8 A regression of betas in the current period, β_2, against betas in the last period, β_1, yields the equation

$$\beta_2 = 0.8\beta_1 + 0.23.$$

If the betas of two stocks in the current period are 0.8 and 1.2, use Blume's technique to predict their betas in the next period.

Question 5.9 A regression of betas in the current period, β_2, against betas in the last period, β_1, yields the equation

$$\beta_2 = 0.9\beta_1 + 0.13.$$

If the betas of three stocks in the current period are 0.8, 1, and 1.2 use Blume's technique to predict their betas in the next period.

Question 5.10 A regression of betas in the current period, β_2, against betas in the last period, β_1, yields the equation

$$\beta_2 = A\beta_1 + B.$$

A manager gives the same data to each of three interns and asks them to compute A and B. The three interns get the following differing results

1. $A = 0.634, B = 0.350$,
2. $A = 1.2, B = -0.2$,
3. $A = 0.7, B = 0.1$.

The manager does not have the time to check the calculations. Which values of A and B does she use, and which intern gets a permanent job? Justify your answer.

If the betas of three stocks in the current period are 0.8, 1, and 1.2 use Blume's technique to predict their betas in the next period. Discuss briefly why Blume's technique is plausible.

6
Multi-factor models

A single-factor model is very simplistic. Do we really believe that all correlation between stocks arises from the level of the market? Stocks in the same sector will have much more correlation than stocks from competing sectors. Stocks from different countries will have less correlation than stocks from the same country. In particular, suppose we take two US oil stocks. These will both be exposed to the state of the US economy and the price of oil and hence we can expect a high degree of correlation. Whereas the stocks are likely to have little in common with a British biscuit manufacturer or an Australian social networking site. To address this incongruity between the model and the observed market, we might introduce more complexity into the modelling by assuming the market is driven by a number of *uncorrelated factors*. For example, one could consider industry sectors or countries.

6.1 Mathematical formulation

We define our multi-factor model using a number of uncorrelated indices, I_j, and set

$$R_i = a_i + \sum_{j=1}^{L} b_{ij} I_j + c_i, \tag{6.1}$$

where the numbers a_i and b_{ij} are constants, and c_i is a random variable with zero mean and uncorrelated with the indices, I_j. As in the single-factor case, c_i expresses the idiosyncratic risk.

Importantly, we make the key simplifying assumption that

$$\mathbb{E}(c_i c_j) = 0$$

for all $i \neq j$. Without this assumption, the model does not reduce complexity and is equivalent to the unrestricted model.

We next explore what properties this multi-factor model has. Let

$$\mathrm{Var}(I_j) = \sigma_j^2, \ \mathrm{Var}(c_i) = \sigma_{c_i}^2.$$

We then find we have the following key theorem.

Theorem 6.1

$$\mathbb{E}(R_i) = a_i + \sum_{j=1}^{L} b_{ij}\mathbb{E}(I_j), \tag{6.2}$$

$$\mathrm{Var}(R_i) = \sum_{j=1}^{L} b_{ij}^2\sigma_j^2 + \sigma_{c_i}^2, \tag{6.3}$$

$$\mathrm{Cov}(R_i, R_j) = \sum_{k=1}^{L} b_{ik}b_{jk}\sigma_k^2. \tag{6.4}$$

The first of these follows from the linearity of expectation. That is,

$$\begin{aligned}\mathbb{E}(R_i) =& \mathbb{E}(a_i) + \sum_{j=1}^{L} \mathbb{E}(b_{ij}I_j) + \mathbb{E}(c_i), \\ =& a_i + \sum_{j=1}^{L} b_{ij}\mathbb{E}(I_j).\end{aligned} \tag{6.5}$$

To find the variance, we need to compute the expectation of $(R_i - \mathbb{E}(R_i))^2$. This is equal to

$$\mathbb{E}\left(\left(c_i + \sum_{j=1}^{L} b_{ij}(I_j - \mathbb{E}(I_j))\right)^2\right).$$

The assumption of zero correlation between the c_i and the indices I_j allows us to discard the cross terms to get the expression

$$\mathbb{E}\left(c_i^2\right) + \sum_{j=1}^{L} b_{ij}^2\mathbb{E}((I_j - \mathbb{E}(I_j))^2) = \sigma_{c_i}^2 + \sum_{j=1}^{L} b_{ij}^2\sigma_j^2.$$

Finally, to derive the covariance formula we need to compute

$$\mathbb{E}((R_i - \mathbb{E}(R_i))(R_j - \mathbb{E}(R_j)).$$

This equals

$$\mathbb{E}\left(\left(c_i + \sum_{k=1}^{L} b_{ik}(I_k - \mathbb{E}(I_k))\right)\left(c_j + \sum_{l=1}^{L} b_{jl}(I_l - \mathbb{E}(I_l))\right)\right).$$

As with the variance, all the cross terms disappear, and we are left with

$$\mathbb{E}\left(\sum_{k=1}^{L} b_{ik}(I_k - \mathbb{E}(I_k))b_{jk}(I_k - \mathbb{E}(I_k))\right),$$

which is equal to

$$\sum_{k=1}^{L} b_{ik}b_{jk}\sigma_k^2.$$

We illustrate this with a simple example of a 2-factor model.

Example 6.2 Suppose in a two-factor model, we have

	A	B
a_i	1	2
b_{i1}	0.7	1
b_{i2}	0.8	1.2
σ_{c_i}	1.9	0.9

and

$$\begin{aligned} \bar{I}_1 &= 7, \\ \bar{I}_2 &= 3, \\ \sigma_{I_1} &= 1.5, \\ \sigma_{I_2} &= 2. \end{aligned}$$

We then compute for each of A and B return and variance, as well as their covariance, as follows.

For A, the expectation is

$$1 + 0.7 \times 7 + 3 \times 0.8 = 8.3,$$

and the variance is

$$0.7^2 \times 1.5^2 + 0.8^2 \times 2^2 + 1.9^2 = 7.2725.$$

For covariance, we get

$$0.7 \times 1 \times 1.5^2 + 0.8 \times 1.2 \times 2^2 = 5.415.$$

Similarly, for the expectation and variance of B. 

Clearly, we need less data than for a general model but more than for a single-factor model. More specifically, the model has

$$2N + 2L + LN$$

parameters to calibrate. If

$$N \sim 500, \qquad \text{and} \qquad L \sim 10,$$

this is certainly more manageable than the general model but much worse than a single-factor model.

6.2 Types of multi-factor models

We could use various different quantities as driving factors for our multi-factor model. These give different types of models including

- sector-based models;
- macro-economic models;
- statistical models.

One could also consider mixed models; that is, models that contain a mixture of the above factor types.

Sector-based models are intuitive and popular. One has an index for the overall market and also one for each industrial sector. Examples of sectors include:

- banks;
- oil;
- pharmaceuticals;
- steel.

One could also extend this model to have factors for each country, as well for each sector. Note that these factors do not fit our model definition, since they are not uncorrelated. However, we will later show in Section 6.3 that it is possible to obtain a set of uncorrelated indices from a correlated set, thereby solving this problem.

These is evidence that the market index accounts for about 30–50% of the variability of stocks, and that introducing industrial sectors explains another 10%.

A second approach uses the fact that stock prices are driven by the wider economy. Factors could therefore include:

- price of oil;
- inflation;
- government bond yields;
- corporate bond yields (or spreads);
- economic growth.

Yet another approach comes at the problem from a completely different angle. Rather than trying to find economically meaningful variables, the method applies statistical techniques to find the factors which explain most. The standard technique is called *principal components analysis.* Essentially this means finding eigenvalues and eigenvectors of the covariance matrix, keeping the two or three largest eigenvalues and then setting the rest to zero.

Whilst these approaches are all valid, and they all have some intellectual appeal, we need to ask ourselves whether multi-factor models are in fact worth the extra effort. How can we answer this question? One way is to observe that they give a better fit to the historical correlation matrix. However, that is not a convincing argument, for if that were our sole criterion, we would simply use the historical correlation matrix.

Perhaps then it would do us good to remember the scientific method, in particular that a theory is assessed by taking the predictions it makes and experimentally testing them. The quality of its assumptions are not what matters.

Let us then evaluate how well the model predicts future correlations. This will determine how good the model is at suggesting good investments when using mean–variance analysis.

Various techniques have been assessed for their ability to predict future correlation matrices in various studies. Curiously, the results in decreasing order of predictive ability, are:

1. Blume's technique for predicting betas in a single-factor model;
2. single-factor model;
3. multi-factor model;
4. historical covariance.

For one study see [5]. These suggest that, for doing mean-variance analysis, one should adopt a single-factor model with Blume's technique for measuring betas rather than the other methods we have discussed.

6.3 Orthogonalisation for multi-factor models

We have assumed in the above that the indices driving our multi-factor model are uncorrelated. This is unrealistic for any natural choice of indices since there will always be some correlation between sectors, countries or economic factors. For example, there is correlation between the US market index and the price of oil.

Fortunately, there is a standard procedure for removing correlation: the Gram–Schmidt orthogonalisation procedure. Gram–Schmidt is an algorithm designed for inner product spaces. We therefore have to phrase our problem in terms of an *inner product*: this is a generalisation of the familiar dot (or scalar) product for vectors in $\mathbb{R}^n$.

We have a set X and a map from pairs of elements of X to $\mathbb{R}$. This map is required to have some of the properties of the dot product. It is typically written using $\langle$ and $\rangle$.

We require

$$\begin{aligned}\langle \lambda X+\mu Y,Z\rangle =&\lambda\langle X,Z\rangle+\mu\langle Y,Z\rangle,\\ \langle X,Y\rangle =&\langle Y,X\rangle.\end{aligned}$$

The requirement that

$$\langle X,X\rangle = 0 \implies X = 0$$

is generally also made.

For $x,y \in \mathbb{R}^n$, we define

$$\langle x,y\rangle = \sum_{j=1}^{n} x_j y_j = x.y = x^{\mathrm{T}} y.$$

It is easy to see that the dot product satisfies the above requirements for an inner product.

Importantly for our work here, we can describe covariance as an inner product. That is, for two random variables, X and Y, we can define

$$\langle X,Y\rangle = \mathrm{Cov}(X,Y) = \mathbb{E}((X-\mathbb{E}(X))(Y-\mathbb{E}(Y))).$$

Certainly, we have the usual properties of inner products:

$$\langle \lambda X+\mu Y,Z\rangle =\lambda\langle X,Z\rangle+\mu\langle Y,Z\rangle, \tag{6.6}$$

$$\langle X,Y\rangle =\langle Y,X\rangle. \tag{6.7}$$

We can define

$$||X||^2 = \langle X,X\rangle = \mathrm{Var}(X). \tag{6.8}$$

However, noting that $\langle X,X\rangle = 0$ if and only if X is constant, it follows that $\langle\cdot,\cdot\rangle$ is a true inner product only if we restrict to random variables with mean zero.

Definition 6.3 If X and Y are elements of an inner product space, they are said to be *orthogonal* if

$$\langle X,Y\rangle = 0.$$

For vectors in $\mathbb{R}^n$ this means that they are at right angles. For random variables with covariance as an inner product, they are orthogonal if and only if they have zero covariance and so zero correlation.

We are now in a position to describe the Gram–Schmidt orthogonalisation algorithm.

Suppose we have random variables X_i, such that for each j, X_j is not a linear combination of the first $j-1$ terms. Then Gram–Schmidt allows us to replace each X_j with Y_j in such a way that the terms Y_j are orthogonal and

$$Y_j = \sum_{i \le j} \lambda_{i,j} X_i,$$

with $\lambda_{j,j} = 1$. We work successively, first finding Y_1 then Y_2 and so on.

First, we set

$$Y_1 = X_1.$$

We then look for α such that

$$\langle X_2 + \alpha Y_1, Y_1 \rangle = 0.$$

Expanding, we have

$$\langle X_2, Y_1 \rangle + \alpha \langle Y_1, Y_1 \rangle = 0, \quad \text{hence} \quad \alpha = -\frac{\langle X_1, X_2 \rangle}{\langle Y_1, Y_1 \rangle}.$$

So we set

$$Y_2 = X_2 + \alpha Y_1 = X_2 + \alpha X_1.$$

More generally, suppose now we have now found $Y_1, Y_2, \ldots, Y_j$ which are orthogonal. How do we find Y_{j+1}? We let

$$Y_{j+1} = X_{j+1} + \sum_{i \le j} \alpha_{j+1,i} Y_i,$$

and solve for the terms $\alpha_{j+1,i}$. For each $r \le j$, we have

$$\langle Y_{j+1}, Y_r \rangle = \langle X_{j+1}, Y_r \rangle + \alpha_{j+1,r} \langle Y_r, Y_r \rangle,$$

since the terms Y_i for all $i \le j$ are all orthogonal to each other. We therefore set

$$\alpha_{j+1,r} = -\frac{\langle X_{j+1}, Y_r \rangle}{\langle Y_r, Y_r \rangle}. \tag{6.9}$$

One now just repeats going through all values of j.

We illustrate this technique with some examples.

Example 6.4 Suppose our covariance matrix is

$$\begin{pmatrix} 1 & 1 & 1 \\ 1 & 2 & 1 \\ 1 & 1 & 3 \end{pmatrix}$$

and we wish to orthogonalise.

We set $Y_1 = X_1$. We now have

$$\langle X_2, Y_1 \rangle = \langle X_2, X_1 \rangle = 1,$$

and

$$\langle Y_1, Y_1 \rangle = \langle X_1, X_1 \rangle = 1.$$

We therefore put

$$\begin{aligned} Y_2 =& X_2 - \langle X_2, Y_1 \rangle \frac{Y_1}{\langle Y_1, Y_1 \rangle}, \\ =& X_2 - X_1. \end{aligned}$$

To continue we need to find $\langle X_3, Y_1 \rangle$ and $\langle X_3, Y_2 \rangle$. We find

$$\langle X_3, Y_1 \rangle = \langle X_3, X_1 \rangle = 1,$$

and

$$\langle X_3, Y_2 \rangle = \langle X_3, X_2 - X_1 \rangle = 1 - 1 = 0.$$

We therefore set

$$Y_3 = X_3 - X_1,$$

and we are done.

Note that, in most cases, we would also have to subtract a multiple of Y_2 from X_3 to get Y_3. This multiple of Y_2 would probably contain elements of X_1 in addition. ◇

We have presented an algorithm for making the indices have zero correlation (and covariance.) We have not placed requirements on the variances of the indices. This is sometimes done to make all variances 1; but this is easy once they are orthogonal – just divide by the standard deviation.

Note that the result of the algorithm depends on the order in which the indices are taken. If we started at the end then we could get a different (but acceptable) answer. We illustrate this with the next example.

Example 6.5 Now suppose our covariance matrix is

$$\begin{pmatrix} 3 & 1 & 1 \\ 1 & 2 & 1 \\ 1 & 1 & 1 \end{pmatrix}.$$

We let

$$Y_1 = X_1,$$

as usual. We then have

$$\langle Y_1, Y_1 \rangle = \langle X_1, X_1 \rangle = 3.$$

We compute

$$\langle X_2, Y_1 \rangle = \langle X_2, X_1 \rangle = 1.$$

So we put

$$Y_2 = X_2 - \frac{Y_1}{\langle Y_1, Y_1 \rangle} = X_2 - \frac{1}{3} X_1.$$

We then have

$$\begin{aligned}
\langle Y_2, Y_2 \rangle &= \left\langle X_2 - \frac{1}{3}X_1, X_2 - \frac{1}{3}X_1 \right\rangle, \\
&= \langle X_2, X_2 \rangle - \frac{2}{3}\langle X_1, X_2 \rangle + \frac{1}{9}\langle X_1, X_1 \rangle, \\
&= 2 - \frac{2}{3} + \frac{1}{3}, \\
&= \frac{5}{3}.
\end{aligned}$$

We also have

$$\langle X_3, Y_1 \rangle = 1,$$

and

$$\begin{aligned}
\langle X_3, Y_2 \rangle &= \left\langle X_3, X_2 - \frac{1}{3}X_1 \right\rangle, \\
&= \langle X_3, X_2 \rangle - \frac{1}{3}\langle X_3, X_1 \rangle, \\
&= \frac{2}{3}.
\end{aligned}$$

So

$$\begin{aligned}
Y_3 &= X_3 - \frac{1}{3}Y_1 - \frac{2}{3}\frac{3}{5}Y_2, \\
&= X_3 - \frac{1}{3}X_1 - \frac{2}{5}\left(X_2 - \frac{1}{3}X_1\right), \\
&= X_3 - \frac{1}{5}X_1 - \frac{2}{5}X_2.
\end{aligned}$$

◇

For the second example, we had

$$\begin{aligned}
Y_1 &= X_1, \\
Y_2 &= X_2 - \frac{1}{3}X_1, \\
Y_3 &= X_3 - \frac{1}{5}X_1 - \frac{2}{5}X_2,
\end{aligned}$$

whereas for the first, we had

$$\begin{aligned} Y_1 &= X_1, \\ Y_2 &= X_2 - X_1, \\ Y_3 &= X_3 - X_1. \end{aligned}$$

The covariance matrices were the same with X_1 and X_3 switched. However, the outputs of the Gram–Schmidt algorithm do not switch in the same way.

6.4 Review

By the end of this chapter, the reader should be able to do the following questions and tasks.

1. State the definition of multi-factor models for stock returns.
2. Derive the expected returns, variances and covariances of returns in multi-factor models for stock returns.
3. What are the advantages of multi-factor models over single-factor models for stock returns?
4. How many parameters are there in a multi-factor model?
5. Give three different ways to choose factors for a multi-factor model.
6. How can we assess which of two models for return covariances is better? Rank the methods of historical covariance, single-factor model, multi-factor model, and single-factor with Blume's technique, with best first.
7. What is an inner product?
8. What properties does an inner product have?
9. Give an example of an inner product.
10. Does covariance define an inner product on the space of random variables?
11. What is the purpose of Gram–Schmidt algorithm?
12. Does the output of the Gram–Schmidt algorithm depend on the order of the inputs?

6.5 Problems

Question 6.1 Suppose we wish to perform mean–variance analysis with five assets, how many different input numbers do we need in each of the following cases:

- a general covariance matrix;

- a single-factor model;
- a multi-factor model with two indices.

Question 6.2 Suppose in a two-factor model, we have

	A	B
a_i	2	3
b_{i1}	0.9	1.1
b_{i2}	0.8	1.3
σ_{c_i}	1.4	1.5

and

$$\bar{I}_1 = 4,\quad \bar{I}_2 = 5,\quad \sigma_{I_1} = 3,\quad \sigma_{I_2} = 4.$$

Compute for each of A and B return and variance, as well as their covariance.

Question 6.3 Suppose in a two-factor model, we have

	A	B
a_i	0	1
b_{i1}	0.5	0.5
b_{i2}	0.6	0.4
σ_{c_i}	0.2	0.2

and

$$\bar{I}_1 = 3,\quad \bar{I}_2 = 4,\quad \sigma_{I_1} = 2,\quad \sigma_{I_2} = 3.$$

Compute for each of A and B return and variance, as well as their covariance.

Question 6.4 Suppose in a two-factor model, we have

	A	B
a_i	1	2
b_{i1}	2	0.2
b_{i2}	0.2	1.1
σ_{c_i}	2	2

and

$$\bar{I}_1 = 6,\quad \bar{I}_2 = 7,\quad \sigma_{I_1} = 4,\quad \sigma_{I_2} = 5.$$

Compute for each of A and B return and variance, as well as their covariance.

Question 6.5 In a multi-factor model with uncorrelated indices

	A	B
a_i	2	6
b_{i1}	2	1
b_{i2}	1	2
σ_{c_i}	1	1

and

$$\bar{I}_1 = 6,\quad \bar{I}_2 = 7,\quad \sigma_{I_1} = 1,\quad \sigma_{I_2} = 1.$$

If the risk-free rate is 1, find the tangent portfolio and minimal variance portfolio.

Question 6.6 In a multi-factor model with two correlated indices, we have

	A	B
a_i	2	6
b_{i1}	2	1
b_{i2}	1	2
σ_{c_i}	1	1

and

$$\begin{aligned}\bar{I}_1 &= 6\\ \bar{I}_2 &= 7\\ \sigma_{I_1} &= 1\\ \sigma_{I_2} &= 1\end{aligned}.$$

If the correlation coefficient of I_1 and I_2 is 0.25, and the risk-free rate is 8, find the tangent portfolio and minimal variance portfolio.

Question 6.7 Suppose a correlation matrix has entries 1 on the diagonal and $\rho > 0$ off it. Find a single-factor model with this correlation matrix.

Question 6.8 Suppose a model has L factors and indices I_j which are uncorrelated. Suppose the return of asset i is

$$\alpha_i + \sum_{j=1}^{L} a_{ij} I_j + e_i,$$

where $\mathbb{E}(I_j) = \beta_j$, e_i has mean 0, is uncorrelated with I_j for all j and the correlation of e_i and e_j is $\rho > 0$. If all random variables are jointly normal, show that it is possible to express the model as having $L+1$ factors and uncorrelated residuals.

Question 6.9 Let the vectors X_1,X_2, and X_3 have the values

$$\begin{pmatrix}1\\0\\1\end{pmatrix}, \begin{pmatrix}1\\2\\3\end{pmatrix}, \begin{pmatrix}1\\4\\1\end{pmatrix}.$$

Find vectors Y_1, Y_2, and Y_3 which are orthogonal and have

$$Y_j = \sum_{i \leq j} \lambda_{i,j} X_j,$$

for $\lambda_{i,j} \in \mathbb{R}$ with $\lambda_{j,j} = 1$ for all j.

Question 6.10 Repeat the previous question with

$$\begin{pmatrix}1\\0\\1\end{pmatrix}, \begin{pmatrix}1\\2\\1\end{pmatrix}, \begin{pmatrix}1\\5\\2\end{pmatrix}.$$

Question 6.11 Let the vectors X_1, X_2, and X_3 have the values

$$\begin{pmatrix} 1 \\ 1 \\ 1 \end{pmatrix}, \begin{pmatrix} 1 \\ 2 \\ 1 \end{pmatrix}, \begin{pmatrix} 1 \\ 0 \\ 0 \end{pmatrix}.$$

Find vectors Y_1, Y_2, and Y_3 which are orthogonal and have

$$Y_j = \sum_{i \leq j} \lambda_{i,j} X_j,$$

for $\lambda_{i,j} \in \mathbb{R}$ with $\lambda_{j,j} = 1$ for all j.

Question 6.12 Suppose we take a correlated index model and use Gram–Schmidt to create an uncorrelated index model. What will happen to the covariances of stock returns in the model?

Question 6.13 Take the space of all polynomials with real coefficients and define

$$\langle f, g \rangle = \int_0^1 f(x) g(x) dx.$$

Will this define an inner product?

Question 6.14 Find functions g_j for $j = 0, 1, 2$, such that

$$g_j = x^j + \sum_{i<j} \alpha_{i,j} x^i$$

and for $i \neq j$

$$\int_0^1 g_i(x) g_j(x) dx = 0.$$

Question 6.15 Find functions g_j for $j = 0, 1, 2$ such that

$$g_j = x^j + \sum_{i<j} \alpha_{i,j} x^i$$

and for $i \neq j$

$$\int_{-1}^1 g_i(x) g_j(x) dx = 0.$$

7
Introducing utility

7.1 Limitations of mean–variance analysis

Up to this point, we have explored mean–variance analysis and have analysed the geometry of the efficient frontier. We have found a technique for obtaining the tangent portfolio from covariance and return information, and we have explored simplifications to the general model through using single- and multi-factor models of covariance structure.

However, even if we accept that mean and variance are the only factors which investors care about, mean–variance analysis does not tell us which portfolio to hold. It only reduces the set of investments worth considering to those which lie on the efficient frontier of the opportunity set. For investments on the efficient frontier it tells us nothing. Since that set is generally a line or a curve, we still do not have enough information to decide between prospective investments.

To help us make this decision, we need to introduce a further concept. We will also see that the assumption that an investor is only interested in mean and variance leads to some odd conclusions.

The need for something more is also illustrated by the following game known as the St Petersburg paradox.

Example 7.1 We play the following game:

- a player tosses a coin until it comes up tails;
- if this requires n throws you receive 2^n roubles.

How much is the right to play the game worth? We attempt to analyse it using expectations. The probability of terminating after exactly n throws is 2^{-n}. The expected pay-off is therefore

$$\sum_{n=1}^{\infty} 2^{-n} 2^{n} = \infty.$$

If all one cares about is expectation then one should be willing to pay an arbitrarily large amount to play this game.

Practical experiments suggest however that people are not willing to pay very much to play: for example, 1.5 roubles is a typical response. Plus there are credit risk issues, in that who could pay out the sum for large n?

It is a little hard to define variance for this game, since the variance is usually defined in terms of expectation. We can, however, compute for a truncated game in which if there are no tails in the first N throws one receives zero.

For the truncated game the expectation is clearly N, and the variance is equal to

$$\sum_{j=1}^{N} 2^{-j}(2^{j})^{2} - N^{2} = \sum_{j=1}^{N} 2^{j} - N^{2},$$

which clearly goes to infinity as N goes to infinity. ◇

How can we explain people's reluctance to pay very much? One explanation is that not much value is ascribed to a very small probability of winning a very large amount of money. Another, related, explanation is that the prospect of receiving two million dollars is not viewed as being twice as good as receiving one million dollars.

We consider another example.

Example 7.2 In this game,

- we toss a coin;
- if the coin comes up heads you get 11 dollars;
- if the coin comes up tails you lose 10 dollars.

Would you play this game? Certainly the expectation is positive (i.e. 0.5). ◇

In the next example, we suppose the gain is scaled up.

Example 7.3 For this game

- we toss a coin;
- if the coin comes up heads you get X dollars;
- if the coin comes up tails you lose 100 dollars.

For what values of X would you play this game? ◇

Before analysing these examples, we develop a framework.

7.2 Defining utility

Utility functions were introduced in the 18th century by Bernouilli as one way of resolving the St Petersburg paradox. The idea is straightforward: we assign to each possible amount of end-of-period wealth, a non-monetary value, the utility, which expresses how keen we are on it. We then work with expected utility instead of expected value.

Utility theory can be summed up as

He who has most when he dies wins.

However, people generally get more joy of spending money than owning it. (There is, of course, some joy in the ability to contemplate a large bank account balance.) One could therefore argue

He who has spent most when he dies wins.

However, since money can be used to buy consumables, utility can be viewed as a proxy for consumption.

Returning to Example 7.1, suppose we analyse the game using a log utility function. That is, the utility ascribed to a value V is $\log(V)$. It follows that the expected utility is

$$\begin{aligned} \mathbb{E}(\log V) &= \sum_{n=1}^{\infty} \log(2^n)2^{-n}, \\ &= \sum_{n=1}^{\infty} n\log(2)2^{-n}. \end{aligned} \tag{7.1}$$

This is finite and the reader is invited to compute it as an exercise. (Hint: sum the series x^{-n} and differentiate.)

Definition 7.4 We define a utility function to be a function from the positive real numbers, $\mathbb{R}^+$, representing total wealth at the end of the period, to the real numbers, $\mathbb{R}$.

It is important to note that utility is always defined in terms of an investor's **total** holdings rather than an individual asset.

We choose investment in the portfolio X over the portfolio Y if

$$\mathbb{E}(U(X)) > \mathbb{E}(U(Y)),$$

and we are then said to *prefer* X to Y.

Here the portfolio X refers to our total wealth if we adopt a certain investment strategy, and portfolio Y again refers to a total wealth under a different strategy.

We are *indifferent* between portfolios X and Y if

$$\mathbb{E}(U(X)) = \mathbb{E}(U(Y)).$$

Indifference means that we do not care which portfolio we invest in, since both make us equally happy on average.

7.3 Properties of utility functions

Investors will generally prefer more to less. The fact they always prefer more than they have is sometimes called *non-satiation.* It follows that

$$X < Y \implies U(X) < U(Y), \tag{7.2}$$

i.e., U is an increasing function. A decreasing U would therefore say that the investor actually wants less money under certain circumstances. As well as being increasing, utility functions should imply risk aversion. How can we quantify risk aversion?

A risk-neutral investor would not care about variance but only about expectation. So if investments X and Y were such that

$$\mathbb{E}(X) = \mathbb{E}(Y),$$

then the investor would be indifferent between them and so we have

$$\mathbb{E}(U(X)) = \mathbb{E}(U(Y)).$$

However, if he were risk averse and X was risky, and Y always paid the same then we would expect

$$\mathbb{E}(U(X)) < \mathbb{E}(U(Y)).$$

Suppose we have $A < Y < B$, and p is such that

$$Y = (1-p)A + pB.$$

Let X pay A with probability $1-p$ and B with probability p. We then have

$$\mathbb{E}(X) = \mathbb{E}(Y) = Y,$$

but X is risky whereas Y is not. A risk-averse investor would therefore choose Y over X. What does this say about the utility function? Certainly, we want

$$\mathbb{E}(U(X)) < \mathbb{E}(U(Y)) = U(Y).$$

This is equivalent to

$$(1-p)U(A) + pU(B) < U(Y).$$

We recall that a chord is the line between two points on a graph. The line through the points

$$(A, U(A)), (B, U(B))$$

is a chord of the graph of the utility function. This line, V, is given by

$$V((1-p)A + pB) = (1-p)U(A) + pU(B) \quad \text{for } p \in [0,1].$$

We have shown in (7.3) that the graph of U will lie above the graph of V between the points A and B. However, this is precisely the definition of a *(strictly) concave* function. We have therefore shown that a risk-averse investor will have a concave utility function.

If the graph of the utility function were a straight line, then we would have

$$\mathbb{E}(U(X)) = \mathbb{E}(U(Y)),$$

and the investor is then said to be *risk neutral.*

If the utility function were convex then we would have

$$\mathbb{E}(U(X)) > \mathbb{E}(U(Y)),$$

and the investor would prefer a risky asset with the same expectation to a non-risky one. Such an investor is said to be *risk seeking.*

Some typical utility functions are

$$\begin{aligned} &\log(x), \text{ log utility,} \\ &1 - e^{-x}, \text{ exponential utility,} \\ &x - bx^2, \text{ with } b > 0. \end{aligned}$$

The last is called *quadratic utility.*

Note that all these functions are concave – it can be shown that strict concavity is implied by having

$$f''(x) < 0$$

for all x which is true in all these cases.

Importantly, we cannot observe utility functions. Utility functions are only useful as a way of deciding preferences between investments. Two utility functions are said to be *equivalent* if they lead to the same decisions. In particular, if U is a utility function and we take

$$V(X) = a + bU(X)$$

with $a, b \in \mathbb{R}$, and $b > 0$, then U and V are equivalent.

To see this, suppose that

$$\mathbb{E}(U(X)) > \mathbb{E}(U(Y)),$$

then

$$a + b\mathbb{E}(U(X)) > a + b\mathbb{E}(U(Y)),$$

and so

$$\mathbb{E}(V(X)) > \mathbb{E}(V(Y)).$$

We conclude that U and V lead to the same investment decisions; that is, they are equivalent.

7.4 Quadratic utility and portfolio theory

In the previous section we noted that a utility function is quadratic if

$$U(X) = X - bX^2.$$

Such a utility function has some curious properties, but it is popular since it is equivalent to saying that only mean and variance are important. It is also flawed: very large values of wealth imply decreasing utility. It can, however, be viewed as an approximation to more sensible functions.

Recall that $\mathrm{Var}(X) = \mathbb{E}(X^2) - \mathbb{E}(X)^2 = \sigma_X^2$, with σ_X the standard deviation. If U is quadratic, we have

$$\begin{aligned}\mathbb{E}(U(X)) &= \mathbb{E}(X - bX^2), \\ &= \mathbb{E}(X) - b\mathbb{E}(X^2), \\ &= \mathbb{E}(X) - b\,\mathrm{Var}(X) - b\mathbb{E}(X)^2, \\ &= \mathbb{E}(X) - b\sigma_X^2 - b\mathbb{E}(X)^2. \qquad (7.3)\end{aligned}$$

This says that the quadratic utility of an investment is determined purely by mean and variance, or equivalently by mean and standard deviation. It follows that choice between portfolios is determined purely by mean and standard deviation.

We have, writing $\mu_X = \mathbb{E}(X)$, that

$$\mathbb{E}(U(X)) = \mu_X - b\mu_X^2 - b\sigma_X^2.$$

We take $b > 0$. It follows that for μ_x very large, utility will decrease as μ_x increases. Clearly, this is not a good model. In particular, the maximum is attained when

$$\mu_X = \frac{1}{2b}.$$

In order for the model to make sense, we will assume that b is small and μ_X is small compared to $1/2b$. In this case, increasing mean will increase utility, as will decreasing variance.

We next consider the relationship between mean–variance analysis and quadratic utility. We start by noting that we previously defined mean–variance efficiency for returns not wealth. Nonetheless, it is not difficult to extend this concept. To see this, suppose we put all of our wealth into a portfolio which returns R for some given period. If we have W_0 initially, we will have

$$W = W_0(1+R),$$

at the end of the specified period.

Since $\mathbb{E}(W) = W_0 + W_0\mathbb{E}(R)$, we see that $\mathbb{E}(R)$ increases if and only if $\mathbb{E}(W)$ increases. Similarly for variance, since $\mathrm{Var}(W) = W_0^2\mathrm{Var}(R)$.

It follows that if a portfolio maximises quadratic utility it must be mean–variance efficient. That is, there is no portfolio with the same variance and higher expectation or the same expectation for lower variance.

7.5 Indifference curves

Utility functions establish preferences between different investments and therefore provide us with a new tool for helping us decide between two different portfolios on the efficient frontier. In other words, we need to find the point on the efficient frontier of greatest utility.

One approach is to plot curves in mean/standard deviation space such that all investments on the curve have the same expected utility. Since expected utility is a function of mean and standard deviation, we can turn the equation for utility around to get standard deviation as a function of mean and expected utility. Varying the mean with fixed expected utility then gives us an *indifference curve.*

Rearranging the equation for expected quadratic utility gives

$$\sigma_X = \sqrt{\frac{\mathbb{E}(X) - b\mathbb{E}(X)^2 - \mathbb{E}(U(X))}{b}}.$$

Figure 7.1 shows an example of utility indifference for quadratic utility with efficient frontier. The mean, $\mathbb{E}(U(X))$, is fixed whilst $\mathbb{E}(X)$ and σ_X vary. We observe that if an indifference curve through a point on the efficient frontier is not tangent to it, then moving along the frontier in some direction will increase utility and hence the point is not optimal. It follows from this simple geometric argument that the portfolio which maximises utility will be the one where the utility indifference curve is tangent to the efficient set.

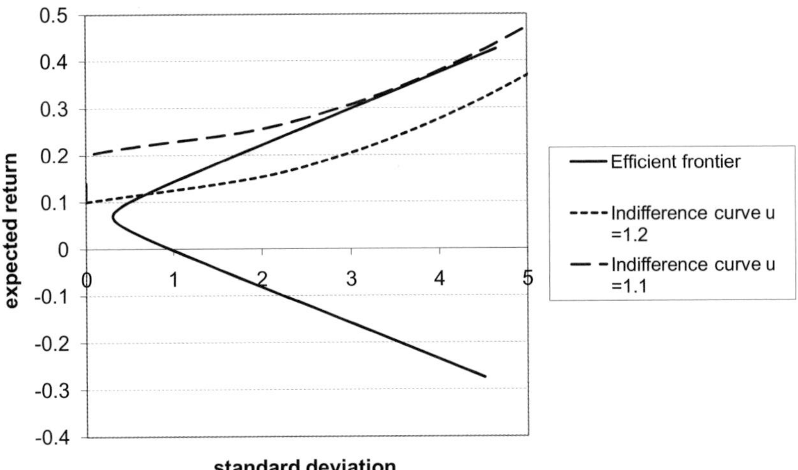

Figure 7.1 Indifference curves and efficient frontier for a mean–variance investor.

7.6 Approximating with quadratic utility

One justification for quadratic utility is that it can be viewed as an approximation to any other utility function.

First, we recall that functions U and V agree to second order at W_0 if

$$U(W) - V(W) = o((W - W_0)^2),$$

where $o((W - W_0)^2)$ means something small compared to $(W - W_0)^2$, i.e.

$$\frac{U(W) - V(W)}{(W - W_0)^2} \to 0$$

as $W \to W_0$.

If U is a general utility function, we can use Taylor's theorem to approximate by a quadratic:

$$U(W) = U(W_0) + U'(W_0)(W - W_0) \\ + U''(W_0)(W - W_0)^2/2 + o((W - W_0)^2).$$

We therefore define

$$M(W) = U(W_0) + U'(W_0)(W - W_0) + U''(W_0)(W - W_0)^2/2. \tag{7.4}$$

to give a quadratic approximation to U.

As long as $W - W_0$ is small the approximation will be good. In other words, provided the change in wealth caused by the investment is small, we can use mean–variance theory.

7.7 Indifference pricing

We can use utility to price. Typically, we have an initial wealth W_0, and we then have a choice between investing in a portfolio which changes our wealth by a random variable X, or by putting it into something worth a fixed amount I. The value of I which makes

$$\mathbb{E}(U(W_0 + X)) = \mathbb{E}(U(I)) = U(I),$$

is the wealth at which the investor is indifferent. The value of X to the investor is then

$$I - W_0$$

and is called the *indifference price* for X. This could be either positive or negative. Note that since U is increasing, it will be invertible and we can write

$$I = U^{-1}(\mathbb{E}(U(W_0 + X))).$$

We summarise the algorithm for computing the indifference price. We suppose have an initial wealth W_0 and an investment that pays X at the end of the period yielding a terminal wealth of $W_0 + X$:

- compute $\mathbb{E}(U(W_0 + X))$;
- let I equal $U^{-1}(\mathbb{E}(U(W_0 + X)))$;
- the indifference price, P, is $I - W_0$.

Example 7.5 If the utility function is linear then

$$U(X) = aX + b, \ \ a > 0,$$

and

$$\mathbb{E}(U(W_0 + X)) = a(W_0 + \mathbb{E}(X)) + b = U(W_0 + \mathbb{E}(X)).$$

So the indifference price is simply $\mathbb{E}(X)$. ◇

For a general utility function, this will not be the case. We define the *risk premium* to be the difference between what a risk-neutral investor (only interested in expected returns) would pay – that is, $\mathbb{E}(X)$ – and the indifference price:

$$\text{risk premium} = \mathbb{E}(X) - P.$$

Example 7.6 Suppose an investor has 100,000 and has a log utility function. Suppose an investment, Y, pays 150, and -50 with probability 0.5. What is the indifference price?

To proceed, we need to know the following:

- the initial wealth;
- the utility function;
- the distribution of the final value of investment.

We have this information and we compute the expectation of $U(Y)$; that is,

$$\mathbb{E}(\log(100,000+Y)).$$

We then exponentiate (i.e., apply U^{-1}). Now

$$\log(100,000+150) = 11.51442434,$$

and

$$\log(100,000-50) = 11.51242534.$$

The expected utility is therefore $\mathbb{E}(U(W)) = 11.51342484$ and the indifference wealth is 100,049.95.

This means that the indifference price is 49.95. ◇

In Example 7.6, $\mathbb{E}(Y)$ is 100,050.00. The indifference price is very similar to the expected value: the risk premium is 5 cents. The investor is only demanding 5 cents to avoid the risk. We note also that the investor's initial wealth was very important in this example. We cannot assess investments without knowing the entire wealth of the investor. We illustrate this in the modified example below.

Example 7.7 Suppose instead the investor's initial wealth was 1,000. We now must compute

$$\mathbb{E}(\log(1,000+Y)),$$

and exponentiate.

$$\log(1,000+150) = 7.047517221,$$

and

$$\log(1,000-50) = 6.856461985.$$

The expected utility is therefore $\mathbb{E}(U(W)) = 6.951989603$. and the indifference wealth is 1045.227248. This implies that the indifference price is 45.23.

The expected return is 1,050 and hence the risk premium has increased to 4.77. A lower initial wealth has resulted in a higher risk premium. ◇

7.8 Review

By the end of this chapter, the reader should be able to do the following questions and tasks.

1. Why is mean–variance analysis not enough to decide between investments?
2. What is the St Petersburg paradox?
3. What properties would you expect a utility function to have and why?
4. What does it mean for two utility functions to be equivalent?
5. Give examples for three typical utility functions.
6. If an investor is risk neutral, what can we say about his utility function?
7. If an investor is risk averse, what can we say about his utility function?
8. Define a quadratic utility function.
9. Discuss the issues with quadratic utility.
10. Relate quadratic utility to mean–variance analysis.
11. Use indifference curves to find the portfolio of maximal utility.
12. Find a quadratic utility function that approximates a general utility function.
13. Define indifference prices and risk premia.
14. Explain how to compute indifference prices.

7.9 Problems

Question 7.1 For each of the following functions on the positive real numbers state whether its graph is convex or whether it's concave.

- $f(x) = x^2$.
- $f(x) = x^3$.
- $f(x) = \sqrt{x+1}$.
- $f(x) = 3x + 2$.

Give brief justification. What characteristics of investors would each function imply if they were used as utility functions?

Question 7.2 Repeat previous question for

- $f(x) = 1 - \exp(-x)$.
- $f(x) = x^5$.
- $f(x) = x^{2/3}$.
- $f(x) = 100x - 500$.

Question 7.3 An investor has the utility function $aX - bX^2$. What is his expected utility if he invests in something that yields terminal wealth $c + dZ$ where Z is a standard $N(0,1)$ random variable.

Question 7.4 State which, if any, of the following utility functions are equivalent:

- $1 - e^{-x}$;
- $\log(x)$;
- $\log(2x)$;
- $x - x^2$;
- $-e^{-x}$;
- $1 - e^{-2x}$.

Question 7.5 An investor has a log utility function. He has \$1,000,000 currently. He can buy investment A that results in him making \$10,000 with probability p and zero otherwise; or putting his money in a riskless asset and getting \$5,000. For what values of p would he buy A?

Question 7.6 Repeat the previous question with utility function $1 - e^{-ax}$, with $a = 10^{-6}$.

Question 7.7 Repeat the previous question with utility function x.

Question 7.8 Assets X, Y, and Z have the following yearly returns that depend on the particular state of the world:

Probability	State	X	Y	Z
0.2	A	4	5	7
0.3	B	7	5	3
0.3	C	0	1	1
0.2	D	8	8	8

Investors must place all of their money, which is \$1,000,000, in precisely one of these assets. For each of the following investors, rank the assets as far as possible:

1. an investor with a log utility function;
2. an investor with the utility function $\sqrt{X}$.

Question 7.9 An investor has a log utility function. His current wealth is \$10,000. An investment pays X with probability 0.75 and $-2X$ with probability 0.25. Compute the indifference price and risk premium when:

1. $X = 10$;

2. $X = 100$
3. $X = 1000$.

Question 7.10 An investor has a log utility function and wealth of one unit. Find the approximating quadratic utility function.

Question 7.11 An investor has a log utility function and wealth of ten units. Find the approximating quadratic utility function.

Question 7.12 An investor has a log utility function and wealth of one hundred units. Find the approximating quadratic utility function.

Question 7.13 An investor has utility function $-e^{-aX}$. His current wealth is \$10,000 and $a = 1/10,000$. An investment pays X with probability 0.75 and $-2X$ with probability 0.25. Compute the indifference price and risk premium when:

1. $X = 10$;
2. $X = 100$;
3. $X = 1000$.

Question 7.14 Repeat previous question with $a = 1/1,000$.

Question 7.15 Repeat previous question with $a = 1/100,000$.

Question 7.16 The current wealth is \$100,000. An investment pays $2X$ with probability 0.8 and $-3X$ with probability 0.2. Compute the indifference price when:

1. $X = 10$;
2. $X = 100$;
3. $X = 1000$;

for the utility function

$$U(W) = -e^{-W/100,000}.$$

Question 7.17 Let $U(W) = ae^{bW}$. If the investor prefers more to less and is risk averse, what can we say about a and b?

Question 7.18 An investor can invest in any of 750 stocks. He can borrow money at a rate of 10%, and does not believe in placing money in riskless assets. If he has a quadratic utility function, describe the geometry of the set of investments, he would consider in:

- the space of portfolio weights;
- expected return/standard deviation space.

Question 7.19 An investor has utility function $-e^{-aX}$. His current wealth is \$10,000 and $a = 1/10,000$. An investment pays $2X$ with probability 0.25 and $-X$ with probability 0.75. Find the quadratic approximating utility function. Use the approximating utility function to compute the indifference price and risk premium when:

1. $X = 10$;
2. $X = 100$;
3. $X = 1000$.

8
Utility and risk aversion

In Chapter 7 we saw that a concave utility function implies risk aversion on the part of the investor. In this chapter, we explore risk aversion in more detail, define the twin concepts of absolute and relative risk aversion, and then use these to calculate the indifference price of an investment.

8.1 Risk aversion and curvature

We know that concavity leads to risk aversion. We also know that replacing the function $U(W)$ by $aU(W)+b$ leads to identical decisions and so identical indifference prices.

This suggests that the second derivative is important. In fact,

$$U'' < 0$$

is equivalent to strict concavity for twice-differentiable functions. We would therefore expect that making U'' more negative would increase risk premia.

Since U and $V = aU+b$ give the same preferences, any attempt to quantify risk aversion must assign the same risk aversion to both these functions.

Differentiating makes the b disappear:

$$V' = (V-b)'.$$

To get rid of the a we take ratios; we will see that it is

$$A(W) = -U''(W)/U'(W)$$

which is important, and this is the same for U and V.

It turns out that the indifference price for an investor with wealth W_0 is largely determined by $A(W_0)$ and the mean and variance of the investment.

8.2 Absolute risk aversion

Suppose an investor has initial wealth W. Suppose further that we have a gamble Z with mean zero and variance σ^2. There is some wealth I so that the investor is indifferent between $W+Z$ and I.

Since the investor is indifferent, we must have

$$\mathbb{E}(U(W+Z)) = \mathbb{E}(U(I)) = U(I). \tag{8.1}$$

Our investor is risk averse so he will prefer less money with certainty to more money with uncertainty, which means

$$\pi = W - I > 0.$$

The sum π is therefore the sum of money that the investor will give up to avoid the gamble Z. We can view π as the cost of a hedge against Z.

The cost of this insurance is a measure of the investor's *absolute risk aversion* at level W. Suppose we expand using Taylor's theorem and see how we can relate π to the derivatives of U.

Using Taylor's theorem, we can write

$$U(W+Z) = U(W) + U'(W)Z + \frac{1}{2}U''(W)Z^2,$$

plus an error that is small compared to Z^2.

Since $\mathbb{E}(Z) = 0$, and so $\text{Var}(Z) = \mathbb{E}(Z^2) = \sigma^2$, this implies, up to a small error,

$$\mathbb{E}(U(W+Z)) = U(W) + \frac{1}{2}U''(W)\sigma^2. \tag{8.2}$$

We also have from Taylor's theorem that

$$U(I) = U(W) - U'(W)\pi,$$

plus an error that is small compared to π. Using (8.1), we have

$$\frac{1}{2}U''(W)\sigma^2 \approx -U'(W)\pi.$$

On rearranging, we conclude

$$\pi = -\frac{1}{2}\frac{U''(W)}{U'(W)}\sigma^2. \tag{8.3}$$

We have shown that for an investment with zero mean the indifference price is approximated by

$$\frac{1}{2}\frac{U''(W)}{U'(W)}\sigma^2.$$

The risk premium for a bet with zero mean is therefore

$$-\frac{1}{2}\frac{U''(W)}{U'(W)}\sigma^2.$$

Since we can write a general investment, X, as

$$X = (X - \mathbb{E}(X)) + \mathbb{E}(X),$$

(that is, as a zero mean investment $Y = X - \mathbb{E}(X)$ plus a zero variance investment), this provides a good approximation to the risk premium in general, and the general indifference price is

$$\mathbb{E}(X) + \frac{1}{2}\frac{U''(W)}{U'(W)}\sigma^2.$$

We give this important ratio of derivatives a name:

Definition 8.1 The *absolute risk aversion* at level W is defined to be

$$A(W) = -\frac{U''(W)}{U'(W)}.$$

Note that if U is increasing and concave, we have $U'' < 0$ and $U' > 0$, so A is positive.

At this juncture, we take stock and summarise some of the key concepts we have introduced and quantified so far in our study of utility.

- Indifference price: the value we would pay for an investment.
- Risk premium: the amount we would deduct from the expected value of an investment in order to get compensation for its riskiness. That is, the difference between the risk-neutral price and the indifference price.
- Absolute risk aversion at a given level of wealth: this tells us how much to multiply the variance of an investment by to get the risk premium.

Absolute risk level is dependent on wealth level. Generally, we would expect the absolute risk aversion to decrease with wealth. A millionaire is less worried about losing \$100 than a student is. We would therefore expect A to be decreasing i.e.,

$$A'(W) < 0.$$

If $A'(W) = 0$, in other words A is constant, then the risk premium does not vary with wealth.

In the following example we calculate absolute risk aversion for a log utility function.

Example 8.2 Suppose we take

$$U(W) = \log W.$$

Then

$$U'(W) = \frac{1}{W},$$
$$U''(W) = -\frac{1}{W^2}.$$

We therefore have

$$A(W) = \frac{1}{W}$$

which is a decreasing function. The risk premium decreases with increasing wealth. ◇

By comparison, in the next example we calculate absolute risk aversion for an exponential utility investor.

Example 8.3 Suppose we take

$$U(W) = 1 - e^{-aW} \text{ with } a > 0.$$

Then

$$U'(W) = ae^{-aW},$$
$$U''(W) = -a^2e^{-aW}.$$

We therefore have

$$A(W) = a$$

which is constant. The risk premium does not vary with the level of wealth. ◇

8.3 Relative risk aversion

It is also useful to think in terms of *relative risk aversion* where the aversion is in terms of *fractions* of current wealth that might be lost instead of absolute dollar amounts, since we might expect such behaviour to vary less with wealth level.

We therefore consider a random variable Z with

$$\mathbb{E}(Z) = 1, \tag{8.4}$$
$$\mathrm{Var}(Z) = \mathbb{E}(Z^2 - 1) = \sigma^2, \tag{8.5}$$

and suppose that all of our wealth will be invested in Z. We should think of σ^2 as being small.

Since all of our wealth is invested in Z, the indifference wealth is then the value I such that

$$\mathbb{E}(U(WZ)) = U(I).$$

The price one would pay to avoid Z is then $W - I$, or as a fraction of wealth, we obtain

$$\pi_R = \frac{W-I}{W}.$$

We use this derive a formula for relative risk aversion. We will proceed similarly to before and equate the initial terms in the Taylor series. That is, we have

$$WZ - W = W(Z-1),$$

so

$$\mathbb{E}(WZ - W) = W\mathbb{E}(Z-1) = 0,$$

and

$$\mathbb{E}((WZ-W)^2) = W^2\mathbb{E}(Z^2-1) = W^2\sigma^2.$$

Up to third-order terms, we therefore have

$$U(WZ) = U(W) + U'(W)(WZ-W) + U''(W)\frac{(WZ-W)^2}{2},$$

and taking expectations, we obtain

$$\mathbb{E}(U(WZ)) = U(W) + \frac{U''(W)W^2}{2}\sigma^2. \tag{8.6}$$

From Taylor's theorem, we also have

$$\begin{aligned} U(I) =& U(W(1-\pi_R)), \\ =& U(W) + U'(W)(-\pi_R W). \end{aligned} \tag{8.7}$$

Equating terms, we obtain

$$\pi_R = -\frac{\sigma^2}{2}\frac{WU''(W)}{U'(W)}.$$

We therefore make the following definition.

Definition 8.4 The *relative risk aversion* for a utility function U is

$$R(W) = -\frac{WU''(W)}{U'(W)}.$$

Note that an investor with constant absolute risk aversion will display increasing relative risk aversion. We can see this intuitively. If an investor has the same absolute risk aversion to losing \$10 for overall wealth of \$100 or wealth of \$1,000,000, then their risk aversion as a percentage of wealth is much higher for the latter case compared with the former.

In Example 8.2 we saw that $A(W) = W^{-1}$ for log utility. It follows that the associated relative risk aversion is

$$R(W) = WA(W) = 1.$$

We have a constant relative risk aversion.

On the other hand, for exponential utility (see Example 8.3) the relative risk aversion is equal to

$$R(W) = WA(W) = aW,$$

and increases with wealth levels.

8.4 Varying the utility function

Suppose we compute the indifference price for an investor for two different utility functions which have the same absolute risk aversion at level W_0. How similar are the prices?

We consider an investor with a log utility function, and another one with an exponential utility function. We take initial wealth $W_0 = \$100,000$.

For the exponential utility investor, we take

$$U(W) = 1 - e^{-aW}$$

with $a = 1/100,000$.

We consider that an investment pays $50 + X$, and $50 - X$ each with probability 0.5.

For the log utility investor we need to compute

$$\mathbb{E}(\log(100,000 + 50 \pm X)),$$

and then exponentiate.

In Example 7.6, we did this for $X = 100$.

One can carry out a similar analysis for exponential utility. We can examine how the price changes with increasing risk by varying X. We present results both for the exponential and log utility, we have put in detailed derivations for the exponential case only. See Tables 8.1 and 8.2.

X	100	200	400	800
average move	50	50	50	50
up wealth	100150	100250	100450	100850
down wealth	99950	99850	99650	99250
utility of up wealth	0.6327	0.6330	0.6338	0.6352
utility of down wealth	0.6319	0.6316	0.6308	0.6294
$\mathbb{E}(U(X))$	0.6323	0.6323	0.6323	0.6323
indifference wealth	100049.95	100049.80	100049.20	100046.80
investment value	49.95	49.80	49.20	46.80
value with log utility	49.95	49.80	49.20	46.80

Table 8.1 *Indifference price as a function of wealth level for low levels of risk.*

X	1600	3200	6400	12800	25600
average	50	50	50	50	50
up wealth	101650	103250	106450	112850	125650
down wealth	98450	96850	93650	87250	74450
utility up	0.6381	0.6439	0.6551	0.6765	0.7154
utility down	0.6264	0.6203	0.6080	0.5821	0.5250
$\mathbb{E}(U(X))$	0.6323	0.6321	0.6316	0.6293	0.6202
indifference W	100037.20	99998.81	99845.34	99233.03	96808.38
value	37.20	–1.19	–154.66	–766.97	–3191.62
value with log	37.21	–1.19	–154.91	–772.17	–3280.60

Table 8.2 *Indifference price as a function of wealth level for high levels of risk.*

We observe that the indifference price decreases as X increases which expresses the increasing risk in the presence of risk aversion. It eventually becomes negative; investors will pay to avoid risk even when the expectation is positive. The two utility functions suggest very similar indifference prices until the risk is very large, since they have been calibrated to give the same risk aversion at the initial wealth level.

This shows that absolute risk aversion largely determines the risk premium and hence the indifference prices, provided the risk is small compared to the portfolio size.

Another possibility for a utility function is a power utility of the form

$$U(W) = -W^{-1/3}.$$

We wish to choose between investments A and B which result in final total wealths as follows:

A		*B*	
probabilities	value	probabilities	value
0.25	4	0.333333	4
0.5	6	0.333333	6.2
0.25	8	0.333333	8

We compute utilities to obtain

A			*B*		
probabilities	value	utility	probabilities	value	utility
0.25	4	–0.62996	0.333333	4	–0.62996
0.5	6	–0.55032	0.333333	6.2	–0.54434
0.25	8	–0.5	0.333333	8	–0.5

which implies

$$\mathbb{E}(U(W)) = -0.55765, \qquad \mathbb{E}(U(W)) = -0.5581,$$

and suggests that the investor would take A.

We can repeat the same analysis replacing the exponent in the power utility with -0.1 :

A			*B*		
probabilities	value	utility	probs	value	utility
0.25	4	–0.8705505	0.3333	4	–0.8705506
0.5	6	–0.8359588	0.3333	6.2	–0.8332221
0.25	8	–0.8122524	0.3333	8	–0.8122523

which yields

$$\mathbb{E}(U(W)) = -0.838680141, \qquad \mathbb{E}(U(W)) = -0.838675052.$$

The investor would now take B.

8.5 St Petersburg revisited

We saw in the famous St Petersburg example that using a utility function made the expected utility finite thereby allowing us to find a price. However, if our utility function is unbounded, you can always find a lottery with the same issues.

Let U be the utility function. Let X_n be such that

$$U(X_n) \geq 2^n. \tag{8.8}$$

If a lottery pays X_n with probability 2^{-n}, then the expected utility is equal to

$$\sum_{j=1}^{\infty} 2^{-n} U(X_n) \geq \sum_{j=1}^{\infty} 2^{-n} 2^n.$$

8.6 Review

By the end of this chapter, the reader should be able to do the following questions and tasks.

1. Define and derive the absolute risk aversion function associated with a utility function.
2. Define and derive the relative risk aversion function associated with a utility function.
3. How do we compute the indifference price given the absolute risk aversion?
4. How do we compute the indifference price given the relative risk aversion?
5. Suppose an investor has constant absolute risk aversion, what does this tell us about his behaviour?

8.7 Problems

Question 8.1 Compute the relative and absolute risk aversions for the following utility functions:

$$-W^{-1/4}, \quad -W^{-1/2}, \quad -W^{-1/3}, \quad 1-e^{-W}.$$

Question 8.2 For each of the following utility functions, compute the absolute and relative risk aversions:

- $U(x) = -1/x$;
- $U(x) = x - ax^2$;
- $U(x) = x$.

Comment on the qualitative features.

Question 8.3 What is meant by the absolute risk aversion of an investor? For each of the following utility functions, compute the absolute risk aversion as a function of wealth:

- $U(W) = W^{1/4}$;
- $U(W) = \log W$;

- $U(W) = -\exp(-W)$.

How do you think the absolute risk aversion should vary with wealth? Justify your answer.

Question 8.4 A rational investor has a \$1,000,000 in wealth. He indicates that he would not enter in an investment with expected value of \$100 and standard deviation of \$100. He also indicates that he would enter in an investment with expected value of \$100 and standard deviation of \$50. For each of the following, state with justification whether it is likely the investor would invest, would likely not invest, or there is too little information to say. All the following investments have expected value \$200.

- standard deviation 50;
- standard deviation 100;
- standard deviation 150;
- standard deviation 200;
- standard deviation 250.

Question 8.5 An investor has a \$1,000,000. He accepts an investment with expected return 50 and standard deviation 100. He rejects an investment with expected return 50 and standard deviation 200. Suppose he has constant relative risk aversion, discuss his likely behaviour as a function of X for an investment of expected return 25 and standard deviation X if his wealth falls to \$100,000.

Question 8.6 Investment strategies A and B have the following returns:

State	A	B
X	10	15
Y	12	10
Z	–10	–15.

The states all occur with probability $1/3$. Investors have a \$1,000,000 and can put all their money in A or B or half their money in each. For each of the following investors, state as much as you can about how they would order the three possibilities:

- an investor with a quadratic utility function;
- an investor with the utility function $\log(W)$;
- an investor with the utility function $5+3\log(W^2)$.

Question 8.7 An investor has \$20,000,000. He accepts an investment with expected return 50 and standard deviation 100. He rejects an investment with expected return 50 and standard deviation 200. His wealth falls to $\$2,000,000$. Discuss his likely behaviour as a function of X for an investment of expected return 50 and standard deviation X if if he has an exponential utility function. Analyse this case also for an investor with a power utility function.

9

Foundations of utility theory

In this chapter we step back and critique utility theory. Certainly, it provides a convenient framework to model investors' choices. However, it is by no means the only way. We may start by asking the following questions in order to assess our model.

- Does utility theory correctly predict an investor's choices?
- Are an investor's choices compatible with utility theory?
- Does utility theory follow from reasonable assumptions?

9.1 Analysing utility theory through experimental economics

We start by considering an example of two bets defined as follows:

Bet A is

- Receive \$11 with probability 0.5.
- Pay \$10 with probability 0.5.

Would you take this?

Bet B is

- Receive \$$X$ with probability 0.5.
- Pay \$100 with probability 0.5.

For what X would you accept B?

In experiments, many people say that they would not take A. Now let us suppose that those who refuse have increasing, concave utility functions. We try to find out what utility theory tells us about X for them. The failure to take bet A says

$$0.5U(W+11)+0.5U(W-10)<U(W),$$

which in turn implies

$$U(W+11)-U(W)<U(W)-U(W-10).$$

There are \$11 from W to $W+11$ and only \$10 from $W-10$ to W. These \$11 going up are worth **less** in *marginal* utility terms than those \$10 going down.

Next we consider the marginal utility (MU) worth of the $W+10$ to $W+11$ dollar. This one will be worth the least, since U is increasing and concave; 11 times its MU is less than the MU of the \$10 going down. The dollar with most MU will be the one at $W-10$. Therefore 10 times its MU is more than the MU of the \$10 going down. Combining these inequalities we have that the $W+11$ dollar's MU is less than $10/11$ of the MU of the $W-10$ dollar.

Let us suppose in addition that the investor's behaviour does not vary much over his wealth changing by \$1000. Therefore, he would not take the bet if his wealth was $W+21$ either. We can repeat the previous argument replacing W by $W+21$. This implies that the MU of the $W+32$ dollar is less than

$$\frac{10}{11}$$

of that of the $W+11$ dollar, and is also less than

$$\left(\frac{10}{11}\right)^2=\frac{100}{121}$$

of the $W-10$ dollar.

More generally, we know that for $W \leq Y \leq W+1000$, the investor values the $Y+11$ dollar less than $10/11$ of the $Y-10$ dollar. Repeating this with $Y=W$, $W+21$, $W+42$, $W+63,\ldots$, we see that the dollar at $W-10+21k$ is worth less than

$$\left(\frac{10}{11}\right)^k$$

of the $W-10$ dollar. For example, setting $k=30$, we have that the $W+920$ dollar has MU less than

$$\left(\frac{10}{11}\right)^{30}\sim 6\%$$

that of the $W-10$ dollar.

Rabin and Thaler [16] have shown that if the investor would not take the first bet for any initial wealth, then he or she would never take the second bet for any value of X. Hence if someone were to turn down the first bet regardless of his or her initial wealth, he or she would turn down a 50–50 chance to win a billion dollars with a stake of only \$100. Furthermore, without making the assumption that the person would behave similarly for varying values of initial wealth, they also show that a person with $340,000 who turns down the first bet at 105 versus 100 would require $5,500,000 to risk losing $10,000.

We then must ask what conclusions may be drawn from Rabin and Thaler's work? Firstly, it would seem that reasonable levels of risk aversion for large sums imply virtually no risk aversion for small sums. Alternatively, risk aversion for small sums of money is **not** implied by expected utility theory. For small sums of money investors are effectively risk neutral. Importantly, those who said "no" to the first bet do not make decisions that can be modelled using utility theory. It follows that whilst utility functions provide us with a useful model they do not explain well people's observed tendency to avoid small risks. That said, any model has its limitations. These do not discredit the model but rather we should bear them in mind when applying it.

9.2 The rational investor

We have seen that utility functions lead to behaviour that could be regarded as counter-intuitive. Yet they permeate the economics and finance literature. Rabin describes his work as a "beating a dead horse," that keeps on coming back from the dead. Why? One explanation is that under certain relatively mild assumptions about investor behaviour, one can prove that investment preferences are equivalent to those given by some utility function. Such an investor is sometimes said to be *rational*. If investors are rational, in this sense, then they behave according to expected utility and would not turn down bet A in Rabin and Thaler's example without having other very strange behaviour.

In order to properly define our concept of rationality, we first introduce some notation, for comparison of preferences. That is, we shall write $A < B$ for the case where investment B is preferred to investment A, and $A > B$ for the converse. We shall write $A \sim B$ for the case where the investor is indifferent. We can also write $A \leq B$ for A being preferred or indifferent to B. We use this ordering of preferences to define the properties which we might require of our rational investor, these being our four "axioms 'of rationality." Throughout this section, we will assume the investment encompasses the entire wealth of the investor. Part of the investment could, however, be held in cash.

The first property we could require is *comparability*. Given two investments, precisely one of

1. $A < B$,
2. $A \sim B$,
3. $A > B$,

should hold. This effectively states that the investor should always be able to express an opinion as to the relative merits of two instruments.

Our second property is *transitivity*: if A is preferred to B and B is preferred to C then A must be preferred to C. We also require that if $A \sim B$ and $B \sim C$ then $A \sim C$. Whilst transitivity appears obvious, it is not difficult to find circumstances where it does not hold. For example, when faced with a gamble, a person may say that they are indifferent between a 50–50 gamble for \$100 or \$0, and receiving 36 dollars. When quizzed again they may say the same thing for \$37. However, clearly they will never be indifferent between \$36 and \$37.

The third property is *independence.* Suppose that an investor is indifferent between investments A and B and suppose we have a third investment C. If we define investment D to be A with probability p, and C otherwise; and suppose investment E to be B with probability p, and C otherwise. Independence states that in this case, the investor should be indifferent between D and E. Analysing the possible outcomes, either the investor receives C in both cases which clearly suggests indifference, or the investor receives one of two investments between which he is indifferent and hence again should be indifferent. It is relatively straightforward to extend this property to a collection of n investments using mathematical induction. That is, if we define investment D to be A_i with probability p_i for $i = 1, \ldots, n$, and investment E to be B_i with probability p_i for $i = 1, \ldots, n$, and the investor is indifferent between each A_i and the corresponding B_i, then the investor will be indifferent between D and E. We leave the proof of this as an exercise for the reader.

Another property sometimes used is *certainty equivalence.* This states that the investor is indifferent between any investment and some cash sum. Roughly stated, this says that every investment has a price.

A more subtle property is the Archimedean axiom, also called the continuity axiom. Given three investments X, Y, and Z, such that $X < Y < Z$, then there exists $\alpha \in [0,1]$, such that $G(X,Z,\alpha)$ which is probability α of X and $1-\alpha$ of Z is equivalent to Y. The property is called the Archimedean axiom by analogy with the natural numbers; it states that given numbers X and Y, with $X < Y$, then if you add 1 to X enough times you get Y. This essentially ensures that Y is finite. If Y is infinity, this would not be true. For us, the Archimedean axiom excludes the possibility of one investment being infinitely better than another. We illustrate this with an example of a preference function which does not satisfy the Archimedean axiom.

Example 9.1 Imagine that we have an investor who is fixated with getting \$1,000,000 for some reason: e.g. for medical bills, to pay a ransom or to buy his (potential) betrothed a diamond ring. He prefers any investment with a non-zero probability of getting a $\$1,000,000$ to one without. Hence if X is \$900,000, Y is \$950,000 and $Z = \$1,000,000$, clearly $X < Y < Z$. However, since any prospect no matter how remote of reaching a million is preferred, we have that $G(X,Z,\alpha)$ will be preferred to Y for all $\alpha \in (0,1)$. ◇

We are now ready to define our concept of rationality.

Definition 9.2 A *rational investor* is one whose preferences satisfy the four principal axioms of utility, namely:

- Comparability;
- Transitivity;
- Independence;
- Certainty equivalence/Archimedean, i.e. continuity;

Note that generally only one of the two versions of the final axiom is assumed and then the other is deduced. We shall assume both versions for simplicity.

9.3 The rational expectations theorem

The rational expectations theorem states that an investor's preferences are given by expected utility if and only if their preferences satisfy the four axioms of comparability, transitivity, independence and certainty equivalence. That expected utility implies the four axioms is straightforward as we show below.

Proof We start by assuming that preferences are given by expected utility. Comparability follows by taking the investment with the higher expected utility. That is, since precisely one of

$$\mathbb{E}(U(A)) < \mathbb{E}(U(B)),$$

$$\mathbb{E}(U(A)) = \mathbb{E}(U(B)),$$

$$\mathbb{E}(U(A)) > \mathbb{E}(U(B)),$$

is true, we also have that precisely one of

$$A < B, \qquad A \sim B, \qquad A > B,$$

is true, and comparability follows.

Suppose next that preferences are given by expected utility and $A < B < C$, then

$$\mathbb{E}((U(A)) < \mathbb{E}(U(B)) \quad \text{and} \quad \mathbb{E}(U(B)) < \mathbb{E}(U(C)),$$

hence

$$\mathbb{E}(U(A)) < \mathbb{E}(U(C))$$

and therefore

$$A < C;$$

that is, transitivity is upheld.

To demonstrate independence, suppose that investments A and B are equivalent and accordingly

$$\mathbb{E}(U(A)) = \mathbb{E}(U(B)).$$

We define D to be A with probability p and C with probability $1-p$; and E to be B with probability p and C with probability $1-p$. Then

$$\begin{aligned}\mathbb{E}(U(D)) =& p\mathbb{E}(U(A)) + (1-p)\mathbb{E}(U(C)),\\ =& p\mathbb{E}(U(B)) + (1-p)\mathbb{E}(U(C)),\\ =& \mathbb{E}(U(E)),\end{aligned}$$

as required.

To prove certainty equivalence we will need to assume that the utility function, U, is increasing and continuous. Given these properties, the function U has an inverse U^{-1}. For an investment A, we set

$$I = U^{-1}(\mathbb{E}(U(A)).$$

We then have

$$\mathbb{E}(U(I)) = U(I) = \mathbb{E}(U(A)),$$

so the investor is indifferent between I and A, as required. The Archimedean axiom also follows easily since it can be reduced to simple statements about the real number system.

The converse, that rationality implies that investor preferences are specified by expected utility for some utility function, is more difficult. We will sketch the proof here under the additional assumption of the Archimedean/continuity axiom.

We assume all prospects lie between two extremes, α and β. We can take $\alpha = 0$ and β very large. The fact that α is less than β implies that certain investments are preferred to others. We also assume that there are only finitely many possible values of terminal wealth. Given that money is discrete – that is, you cannot own a tenth of a cent – this is not much of a restriction.

Suppose next we have a gamble $g(k)$ which pays

$$\begin{cases} \beta \text{ with probability } k, \\ 0 \text{ with probability } 1-k. \end{cases}$$

By certainty equivalence, $g(k)$ is worth a sum of money $m(k)$.

Given a fixed sum of money X, it follows from continuity and the fact that all money lies between 0 and β, that X must be equivalent to a gamble $g(k_X)$ for some value of k_X. To see this consider that as k goes to 1, $g(k)$ will approach β and as it goes to zero, $g(k)$ will approach zero. As k varies between 0 and 1, $g(k)$ will pass through all values zero and β.

Our strategy is to show that every security is equivalent to such a gamble. To this end, suppose we have an arbitrary product, P, which pays

$$X_i \text{ with probability } q_i,$$

and $\sum_i q_i = 1$. We note that for each value of X_i there is a gamble $g(k_{X_i})$ that is worth X_i. Now, consider another security, R, that pays

$$g(k_{X_i}) \text{ with probability } q_i.$$

The investment R is therefore a *compound gamble*: we receive the pay-off of another gamble.

By invoking the independence axiom (in the n-asset case) we have that the securities P and R are equivalent. We further note that the security R only ever pays 0 or β and that the probability it pays β is

$$k = \sum q_i k_{X_i}.$$

This means that R is identical to the security $g(k)$ and hence P is equivalent to a gamble $g(k)$.

We next define for a wealth X, the utility of X via $U(X) = k_X$. It follows that the expected utility of P is then

$$\sum_i q_i k_{X_i} = \mathbb{E}(U(X)).$$

We showed that, for every product P, the investor is indifferent to a security that was of the form $g(k_P)$. Assuming that

$$k_1 > k_2 \implies g(k_1) > g(k_2),$$

which is not obvious from our axioms and we really need an extra axiom for this, we get

$$\mathbb{E}(U(P_1)) > \mathbb{E}(U(P_2)) \implies P_1 > P_2.$$

This demonstrates that we get the actual preference from this utility function. This completes our sketchy proof. It is also possible to prove the result under fewer hypotheses. □

The theorem was originally proved by von Neumann and Morgenstern, see [14].

Example 9.3 Consider the four following gambles which pay off a number of millions of dollars according to which of the world states X, Y and Z occurs.

World State	probability	A	B	C	D
X	0.8	1	1	0	0
Y	0.02	1	0	1	0
Z	0.18	1	5	1	5

An investor has $\$10,000,000$. He states that given a choice between A and B, he would choose A. He also states that given a choice between C and D, he would choose D. Are his choices compatible with rational decision making?

No, they are not. The behaviour fails independence. To see this, consider an asset E that pays 1 in state X and zero otherwise. Independence says that if D is better than C then $B = D + E$ is better than $A = C + E$. However, our investor prefers A to B. The fact that investors often exhibit this sort of behaviour is sometimes called the Allais paradox. ◇

Example 9.4 An investor is believed to have $\$5,000,000$. He is believed to be rational in the sense of the rational expectations theorem. Assets X, Y and Z have expected value 100 and standard deviation $1,000$. His indifference price for A is known to be 95. If the pairwise correlations of future prices are 0.5, estimate his indifference price for the following portfolios:

- a portfolio consisting of 0.5 units of X and 0.5 units of Y;
- a portfolio consisting of 1 unit of each of X, Y and Z.

Now suppose it is discovered that the original estimate of his wealth was incorrect and he actually has $\$10,000,000$. How would that affect your answers?

Since he is rational, we can assume that he makes decisions according to expected utility. We can compute from his indifference price using

$$95 = \text{price} = \mathbb{E}(X) - \frac{1}{2}A \times \text{Var}(X)$$

that his absolute risk aversion, A, for his initial wealth is 0.00001.

The first portfolio has expected value 100 and variance $750,000$. Using the same relation, his approximate indifference price is therefore 96.25. His second portfolio has expected value 300 and variance 6,000,000. The approximate indifference price is therefore 270.00.

We did not actually use the initial wealth level in our solution, so changing our estimate of it has no effect. ◇

9.4 Review

By the end of this chapter, the reader should be able to do the following questions and tasks.

1. What does utility theory imply about risk aversion for small sums of money?
2. What axioms do a rational investor's behaviour satisfy?
3. What does the rational expectations theorem say?
4. What does the axiom of comparability say? Show that an investor deciding according to expected utility satisfies this axiom.
5. What does the axiom of transitivity say? Show that an investor deciding according to expected utility satisfies this axiom.
6. What does the axiom of independence say? Show that an investor deciding according to expected utility satisfies this axiom.

9.5 Problems

Question 9.1 Suppose an investor with increasing concave utility function will not take a bet of win 6, lose 5, with equal chances even if his wealth increases by 1,000. What can we say (if anything) about the following bets?

1. Win 12 or lose 10. Equal chances.
2. Win 6 or lose 6.

Question 9.2 Investors A and B make their investment decisions as follows:

- A chooses a portfolio purely on the basis of maximising expected return;
- B chooses a portfolio purely to minimise variance of return.

Discuss whether A and B are rational investors in the sense of utility theory.

10
Maximising long-term growth

In our journey so far we have examined a couple of ways of assessing investments, namely mean–variance analysis and expected utility. Mean–variance analysis gives us a way of comparing investments but is not always able to decide between them. Expected utility maximisation provides an effective way of deciding between investments but, as we have seen in Chapter 9, requires additional assumptions as to the rationality of investor preferences. Here we look at a third approach where we adopt as our criterion the requirement that the investment should do best in the very long run. In other words, we want to maximise the expected long-term growth rate. The crucial phrase is *long-term* rate. We are not looking to win for any fixed time-horizon, rather we want to adopt a strategy that will do best if we wait for an *arbitrarily long* period of time. We go on to show that this is equivalent to maximising log utility and accordingly from our work in Chapter 9 such an investor would satisfy the axioms of rationality.

10.1 Geometric means

We formalise the problem of long-term growth maximisation by assuming that each year we will invest entirely in portfolio that returns a random variable r_j. These returns are assumed to have the **same** distribution each year and to be independent of each other (sometimes referred to as *independent identically distributed or "iid"* returns). If we start with 1, our wealth after N years is therefore given by

$$(1+r_1)(1+r_2)\cdots(1+r_N).$$

The expected value after N years is therefore

$$\mathbb{E}((1+r_1)(1+r_2)\cdots(1+r_N))$$

and the average annual rate of growth is

$$r_g = ((1+r_1)(1+r_2)\cdots(1+r_N))^{1/N} - 1,$$

where clearly, we have

$$(1+r_g)^N = (1+r_1)\cdots(1+r_N).$$

We can identify $1+r_g$, as the *geometric mean* of the numbers $1+r_j$.

We note that there is no requirement for returns to be measured over periods of a year and so we may then refer to more general *periods*.

Suppose next (for simplicity) that the return can take only a finite discrete set of values. For N very large, the fraction of times each value is taken will be its probability. Hence if the possible values are

$$s_j \text{ with probability } p_j \text{ for } j = 1,\ldots,k,$$

then for N large we have that the total growth will converge to

$$(1+s_1)^{Np_1}(1+s_2)^{Np_2}\cdots(1+s_k)^{Np_k}.$$

To get the annualised growth we take the $1/N$th power and subtract 1, which will in turn converge to

$$(1+s_1)^{p_1}(1+s_2)^{p_2}\cdots(1+s_k)^{p_k} - 1.$$

It is this quantity that we seek to maximise. This is made simpler by re-expressing the maximisation problem using logarithms. Since log is strictly increasing it is enough to maximise

$$\begin{aligned}\log((1+s_1)^{p_1}(1+s_2)^{p_2}\cdots(1+s_k)^{p_k}) &= \sum_{i=1}^{k} p_i \log(1+s_i), \\ &= \mathbb{E}(\log(1+r)). \qquad (10.1)\end{aligned}$$

In other words, we have shown that to maximise the long-term growth rate, we must find the portfolio that maximises

$$\mathbb{E}(\log(1+r)),$$

from which we derive a long-term growth rate of

$$e^{\mathbb{E}(\log(1+r))} - 1.$$

It is important to realise that this portfolio need not be mean–variance efficient nor utility maximising, and generally will be neither.

Next suppose that we have a log utility function and our initial wealth is X Our expected utility at the end of the year will be

$$\begin{aligned}\mathbb{E}(\log(X(1+r))) &= \mathbb{E}(\log(X)) + \mathbb{E}(\log(1+r)),\\ &= \log(X) + \mathbb{E}(\log(1+r)).\end{aligned}$$

We conclude that maximising the log utility is the same as maximising the geometric mean, and hence long-term growth. It is important to recognise that the geometric mean is different from the arithmetic mean which we use when computing expected return and that maximising it will give different answers. One can however prove that the geometric mean is always less than or equal to the arithmetic mean. We conclude this section with an example.

Example 10.1 Suppose we have an asset with return distributed as shown below and our problem is to decide what proportion to put into it and what proportion to put into cash.

return R	–0.2	–0.1	0	0.1	0.2
probabilities	0.1	0.2	0.3	0.3	0.1

Let us assume that there are no short-selling restrictions and that there is no interest payable. We put X in the asset and $1-X$ in cash. The return is then simply X times the return of the asset. We use the same proportions for every period. We tabulate the returns for each R above for some differing Xs.

probabilities	0.1	0.2	0.3	0.3	0.1
X	returns				
0.5	–0.1	–0.05	0	0.05	0.1
0.76	–0.152	–0.076	0	0.076	0.152
1	–0.2	–0.1	0	0.1	0.2
2	–0.4	–0.2	0	0.2	0.4

We next compute the log returns from the above table, by taking $\log(1+R)$ for each entry.

X	$\log(1+R)$ for varying R				
0.5	–0.1054	–0.0513	0.0000	0.0488	0.0953
0.76	–0.1649	–0.0790	0.0000	0.0733	0.1415
1	–0.2231	–0.1054	0.0000	0.0953	0.1823
2	–0.5108	–0.2231	0.0000	0.1823	0.3365

We compute to get

X	$\mathbb{E}(\log(1+R))$	long-term growth rate	expected return
0.5	0.003373	0.003379	0.005
0.76	0.003829	0.003836	0.0076
1	0.003439	0.003445	0.01
2	–0.007368	–0.007341	0.02

Analysing these results, we see that the bigger X is, the more we short-sell and the higher the expected return. However, the average long-term growth rate is maximised when $X = 0.76$. By comparison when $X = 2.00$, the long-term growth rate turns negative and we will *eventually* end up down a lot of money. Markowitz [11] simulated this example and found that after 1000 investment periods, \$1 turned into

X	0.5	0.76	1	2
money	54.6	116.98	103.54	0.005

and, as predicted, $X = 2$ did very badly. This was, however, just one path. To find the X that maximises long-term return one could either plot a graph of log return against X or develop an expression and differentiate. ◇

10.2 Kelly's theorem

In [9] Kelly proved the following general result.

Theorem 10.2 *Suppose we are given two investment strategies with annual return rates r and s. Let the wealth after j years be W_j^r, W_j^s. Assume*

$$\mathbb{E}(\log(1+r)) > \mathbb{E}(\log(1+s));$$

then with probability 1 *there will be an N such that*

$$j > N \implies W_j^r > W_j^s.$$

Kelly's theorem says that if you wait long enough, the investment with higher expected log return will win. However, the theorem does not say anything about N, in particular N will vary probabilistically, i.e., N is a random variable, and will not be bounded above. Therefore whilst if you adopt r, you will win, you may have to wait an arbitrarily long time to see your winnings.

Note that the order of statements in the theorem is very important. That is, to say that *with probability* 1 *there exists N such that for $j > N$ …* is quite different from saying *there exists N such that for $j > N$ with probability* 1 …

In the first case, N will be a random variable varying probabilistically, but in the second, there is only one N and it is always the same.

Samuelson [19] objected to the Kelly argument in the following way. The argument that we should use the geometric mean relied on the law of large numbers: with probability 1, the fraction of draws that take a given value converge to the probability of that value, as N tends to infinity. This is purely a statement about behaviour at infinity, **not** about any finite N. In particular, Samuelson argued that, for a given utility function, even if N is very large the log maximisation may do poorly in that function's terms, and, in particular, that the optimal portfolio as a function of N need not converge to the geometric mean maximiser.

Suppose then that we fix a finite N and a very simple utility function:

$$U(X) = X.$$

Returns from year to year should be independent (given a reasonable level of market efficiency,) and hence our wealth after N years will be $(1+r_1)(1+r_2)\cdots(1+r_N)$, with each r_j distributed the same as r and independent. Since the random variables are independent, the expectation is

$$(\mathbb{E}(1+r))^N.$$

This means that to maximise expected wealth, we should maximise

$$\mathbb{E}(1+r),$$

rather than $\mathbb{E}(\log(1+r))$. However, since one statement deals with a fixed time horizon, and the other with behaviour at infinity, they are not contradictory. We have a paradox: Kelly says that to maximise long-term gains, we must maximise $\mathbb{E}(\log(1+r))$; whereas Samuelson says to maximise expected utility for any fixed time horizon, we should maximise $\mathbb{E}(R)$.

We illustrate the issues with an extreme example.

Example 10.3 To take an example of Markowitz, [11], suppose we have a choice of two investments. Investment r has a fixed return of 1% and investment s has a return of -100% with probability 0.5 (i.e, lose all money) and 300% with probability 0.5. We compute:

$$\begin{aligned}
\mathbb{E}(1+r) &= 1+r = 1.01, \\
\mathbb{E}(\log(1+r)) &= \log(1+r) = 0.00995, \\
\mathbb{E}(1+s) &= 0.5 \times 4 = 2, \\
\mathbb{E}(\log(1+s)) &= -\infty.
\end{aligned}$$

After N years, the expected values are

$$1.01^N, \qquad 2^N,$$

whilst the actual values are 1.01^N and

$$4^N \text{ with probability } 2^{-N}, \tag{10.2}$$

$$0 \text{ with probability } 1 - 2^{-N}. \tag{10.3}$$

It follows that if we wait long enough, investment s will have value zero with probability 1, but for any fixed time horizon, it will win in expectation terms. Importantly, we note that r wins with probability 1 but not with certainty; it is possible to get an infinite string of heads when tossing a coin, but it will only happen with probability 0. ◇

We have seen that the geometric mean maximiser wins in the *long term.* In the long term, we are all dead. For a fixed time horizon, it may (and generally does) make more sense to take into account the expected value and its variance. Markowitz, [11], suggests that one should use the geometric mean portfolio as a cap on variance: one should not invest in a mean–variance efficient portfolio whose expected return and variance are higher than those of the geometric mean portfolio. He does not rule out investing in a less risky one, however.

We conclude this section with some typical examples.

Example 10.4 We use geometric means to decide which of the following investments is long-term preferable:

A		*B*	
probability	return	probability	return
0.1	4%	0.1	4%
0.4	6%	0.3	5%
0.4	8%	0.3	6%
0.1	10%	0.2	7%
		0.1	8%

We compute the geometric mean for A as follows:

a	b	c	d
probability	r	$\log(1+r)$	$a \times c$
0.1	4%	0.0392	0.0039
0.4	6%	0.0583	0.0233
0.4	8%	0.0770	0.0308
0.1	10%	0.0953	0.0095
		sum	0.0675

Finally, take exp(sum) − 1 to see that the geometric mean is 0.0699.
Computing similarly for B we have:

a	b	c	d
probability	r	$\log(1+r)$	$a\times c$
0.1	4%	0.03922	0.0039
0.3	5%	0.04879	0.0146
0.3	6%	0.05827	0.0175
0.2	7%	0.06766	0.0135
0.1	8%	0.07696	0.0077
		sum	0.0573

Again we take exp(sum) − 1 to see that the geometric mean is 0.0589. We conclude that investment A is preferred to investment B, since its geometric mean is higher. We also note that we do not really need to do the last part of the calculations, since taking exp(sum) − 1 will preserve ordering. We also note in this case the ordinary (arithmetic) means are 7% and 5.90% and hence the ordering is the same. ◇

We can also apply this sort of analysis to decide how much to leverage up. For example, suppose an investment has a 50–50 chance of a 9% gain or a 5% loss. The expected return is then $0.5(9\% - 5\%) = 2\%$. Suppose in addition that we can leverage by borrowing as much as we want at a zero interest rate. How much should we borrow to maximise long-term growth? We analyse by considering x units allocated to the investment with the remaining $1-x$ to be held cash. At the end of the year we find the return is

$$(1+r)x+1-x=1+rx.$$

The expected return is therefore $0.02x$. We compute $\mathbb{E}(\log(1+r))$ to get

$$0.5\times\log(1+0.09x)+0.5\times\log(1-0.05x).$$

From the graph in Figure 10.1 we see that the maximum occurs between 4 and 5. We find it by differentiation, where the derivative is

$$0.5\left(\frac{0.09}{1+0.09x}-\frac{0.05}{1-0.05x}\right).$$

This equals zero if and only if

$$9(1-0.05x)=5(1+0.09x),$$

which has the unique solution $x=4.444$. A more realistic example might have different borrowing and lending rates.

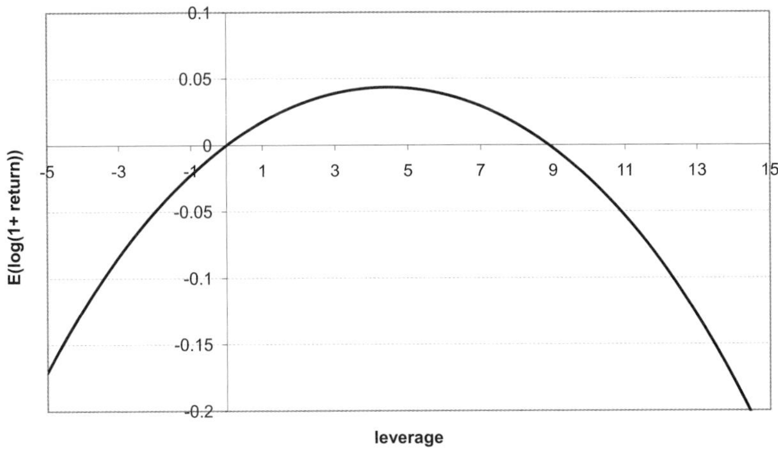

Figure 10.1 Expected log return for different amounts of leverage.

Example 10.5 An investor can borrow unlimited amounts at a rate of 9 and lend at a rate of 5. His current wealth is one million. He can invest unlimited amounts in an asset, X, that returns 20 with probability 0.5 and returns 0 otherwise. If his objective is to maximise wealth in the very long term, how many units of X should he purchase?

If we invest all in X, we get return 20 or 0 with equal probability.

If we invest $\theta < 1$ in X, then the return is $20\theta + (1-\theta)5$ or $(1-\theta)5$ with equal probability.

If we invest $\theta > 1$ in X, then the return is $20\theta + (1-\theta)9$ or $(1-\theta)9$ with equal probability.

We have to maximise $\mathbb{E}(\log(1+r))$. It is differentiable on the range $(0,1)$ and above 1. The expected value of $\log(1+r)$ at 0 is 0.04879 and at 1 is 0.091161.

For $\theta > 1$, we differentiate

$$\frac{1}{2}(\log(1+0.2\theta+(1-\theta)0.09)+\log(1+(1-\theta)0.09)).$$

We then must solve

$$\frac{0.11}{1.09+0.11\theta} - \frac{0.09}{1.09-0.09\theta} = 0.$$

We solve for θ and obtain 1.10101. It is easily checked that there are no critical points in the range $(0,1)$. This means that the maximum must occur at $\theta = 0, 1$

or 1.10101. It is easily checked that the last gives 0.091203 and so gives the greatest value. It is thus the global maximum. ◇

Next suppose we have two investments X and Y such that X has a non-zero probability of losing everything and Y does not. Intuitively, Y should be better than X in the very long term since eventually X will hit zero. Analysing using geometric means, we have for X that

$$\mathbb{E}(\log(1+r_X)) = -\infty,$$

and so the geometric mean is

$$e^{-\infty} - 1 = 0 - 1 = -1.$$

However, the geometric mean of Y will be greater than -1 since

$$\mathbb{E}(\log(1+r_Y)) > -\infty.$$

As we would expect, Y wins in the very long term.

The geometric mean approach tells us that if we want to win in the long term then we should take a fair amount of risk but not so much that we stand to lose everything. Most investors would regard the level of riskiness predicted by geometric mean maximisation as being high although it is guaranteed to win eventually. It is fair to say that most investors are interested in that which occurs in their lifetimes rather than eternity.

10.3 Review

By the end of this chapter, the reader should be able to do the following questions and tasks.

1. Define the geometric mean return of an asset.
2. What does Kelly's theorem say?
3. Which utility function is equivalent to maximising long-term utility?
4. If we have iid returns every year and we want to maximise expected return for precisely 1,000 years in the future what quantity should be maximise?
5. If we have iid returns every year and we want to maximise returns in the very long term, what quantity should we maximise?

10.4 Problems

Question 10.1 Rank the following investments using geometric means.

P return	probability		*Q* return	probability
1%	0.25		2%	0.2
4%	0.5		4%	0.5
9%	0.25		7%	0.3

Question 10.2 Rank the following investments using geometric means.

P return	probability		*Q* return	probability
2%	0.25		2%	0.2
5%	0.5		4%	0.5
6%	0.25		7%	0.3

Question 10.3 Rank the following investments using geometric means.

P return	probability		*Q* return	probability
2%	0.1		2%	0.2
5%	0.4		4%	0.5
7%	0.5		7%	0.3

Question 10.4 Consider the probabilities for rates of return on investments X, Y and Z below. Assuming iid returns, rank them for a very long term investor and for an investor with a fixed time horizon of 100 years who wishes to maximise expected returns.

p	R_X
0.2	6%
0.4	8%
0.3	9%
0.1	11%

p	R_Y
0.1	7%
0.3	8%
0.2	9%
0.3	10%
0.1	11%

p	R_Z
0.4	8%
0.3	9%
0.2	10%
0.1	11%

Question 10.5 Suppose r is such that losing all your money is possible, i.e. a return of -100% has non-zero probability, whereas for s it is not. What can we say about their geometric means?

Question 10.6 An investor can invest or borrow in cash with a return of 2%, or invest in a stock with return of 10% with probability 0.6 and −5% with probability 0.4. How many units of stock should the investor hold if he wishes to maximise his long-term growth rate? What about his expected holdings for a specific date?

Question 10.7 An investor can invest in cash with a return of 2%, borrow with return rate 5%, or invest in a stock with return of 25% with probability 0.5 and -10% with probability 0.5. How many units of stock should the investor hold if he wishes to maximise his long-term growth rate? What about his expected holdings for a specific date?

11
Stochastic dominance

11.1 Introduction

We have developed various methods of comparing investments. These include

- mean–variance efficiency,
- expected utility,
- geometric means.

There are other methodologies. In this chapter, we develop another approach which requires only very weak assumptions on the investor, but strong assumptions about the investments. Part of our objective is to escape from the fact that mean–variance analysis penalises upside, as well as downside, variance. After all, it would be rare for an investor to complain of an unexpected windfall gain.

11.2 Dominance

We now introduce our new key concept as follows; suppose we have portfolios with returns X and Y with the same initial value and suppose that always

$$X \leq Y,$$

at the end of the investment period. One would never prefer X to Y. We can say that Y is *dominant* to X.

Suppose that we build on this with the additional hypothesis that

$$\mathbb{P}(X < Y) > 0.$$

Assuming the investor prefers more to less then clearly an investor would prefer Y to X. We then say that Y is *strictly dominant* to X. We illustrate this in the following example of two hypothetical airlines the performance returns from which vary with the price of oil.

Example 11.1 Consider the airlines JetSafe and BudgetSky. Both have returns that depend on the price of oil.

Price of oil	JetSafe Return	BudgetSky Return
10	15	15
30	14	14
50	13	12
70	12	10
90	10	10

JetSafe is strictly dominant to BudgetSky. Importantly, we note that dominance will occur whatever probability we assign to each market condition. We note, however, that in reality, this situation is not very likely to occur since investors would simply shun BudgetSky and buy JetSafe until the relationship disappeared. (As the purchase price of JetSafe increases from demand, the return would decrease conversely.) ◇

We might ask next how does this new concept of dominance relate to our more familiar notion of efficiency. The answer, is however, that it does not relate particularly well. Whilst an investment Y may be preferred to X, it need not be more efficient than X. Again, we illustrate by way of a simple (and extreme) example.

Example 11.2 Suppose we have investments X and Y defined as follows:

- X returns 0 always;
- Y returns 0 with probability 0.99;
- Y returns 100 with probability 0.01.

Clearly Y is strictly dominant to X. Investment Y is higher in both mean and variance, so efficiency tells us nothing. Furthermore, the semi-variance $S(Y)$ is also greater than 0 so switching to semi-variance does not add anything to our analysis. ◇

11.3 First-order stochastic dominance

We next introduce the more sophisticated idea of stochastic dominance. Firstly, suppose we have two portfolios with returns Y and Z. Suppose they have the same cumulative distribution functions for their returns, i.e., for all real numbers a,

$$\mathbb{P}(Y \leq a) = \mathbb{P}(Z \leq a).$$

We would not be able to distinguish between them using any of our methodologies, since they all are based purely on functionals of our estimates of their probability distributions. We would be indifferent between them.

If Y is strictly dominant to X, and we are indifferent between Y and Z, then we should prefer Z to X. This is the fundamental idea behind stochastic dominance. We do not even need Y to exist, merely that such a hypothetical Y *could* exist is enough. We formalise this with the following definition.

Definition 11.3 We shall say that Z has first-order stochastic dominance over X if

$$\begin{aligned} \mathbb{P}(X \leq a) &\geq \mathbb{P}(Z \leq a), && \text{for all } a, \\ \mathbb{P}(X \leq b) &> \mathbb{P}(Z \leq b) && \text{for some } b. \end{aligned}$$

Importantly, we note that this says nothing about

$$\mathbb{P}(X \leq Z).$$

We illustrate this by expanding on the earlier Example 11.2.

Example 11.4 Suppose we have the following data for the airline JetSafe together with an oil company PetroGiant.

Probability	Price of oil	JetSafe	PetroGiant
0.2	10	15	10
0.2	30	14	10
0.2	50	13	12
0.2	70	12	14
0.2	90	10	15

Clearly, there is no simple relationship between the return on the two companies, but PetroGiant has the same probability distribution of returns as BudgetSky. We next build cumulative distributions as follows.

Return	JetSafe	PetroGiant
9	0	0
10	0.2	0.4
11	0.2	0.4
12	0.4	0.6
13	0.6	0.6
14	0.8	0.8
15	1	1
16	1	1

We see that the cumulative distribution function of PetroGiant is always at least as big as JetSafe's and sometimes bigger. We have first-order stochastic dominance. Buy JetSafe! ◇

We recall that our argument that first-order stochastic dominance implies preferability of an investment was based purely on the investor's desire to have more rather than less. This in turn suggests that any investor with an increasing utility function should prefer a stochastically dominant investment. We formalise this in the following theorem.

Theorem 11.5 *If portfolios X and Y have returns r_X and r_Y and the investor has a utility function U with $U'(s) > 0$ for all s, and the cumulative probabilities satisfy*

$$\begin{aligned}\mathbb{P}(r_X \leq a) &\leq \mathbb{P}(r_Y \leq a) \quad \text{for all } a,\\ \mathbb{P}(r_X \leq b) &< \mathbb{P}(r_Y \leq b) \quad \text{for some } b,\end{aligned}$$

then X will be preferred to Y.

Proof We start by simplifying our notation by writing

$$\begin{aligned}F_X(a) =& \mathbb{P}(r_X \leq a),\\ F_Y(a) =& \mathbb{P}(r_Y \leq a).\end{aligned}$$

We will further assume that the probability distributions are continuous and

$$F_X(a) = \int_{-\infty}^{a} f_X(s)ds.$$

Accordingly,

$$f_X(a) = F_X'(a),$$

and f_X is the density of r_X. We make the analogous assumptions for Y.

For simplicity, we include the additional assumption that there exists some number K for which $f_X(s) = f_Y(s) = 0$, for $|s| \geq K$, and hence

$$F_X(-K) = F_Y(-K) = 0,$$

and

$$F_X(K) = F_Y(K) = 1.$$

We will use integration by parts, for any u, v:

$$\int_a^b u(s)v'(s)ds = u(b)v(b) - u(a)v(a) - \int_a^b u'(s)v(s)ds.$$

Suppose next that the initial wealth is W. If we invest all of our money into portfolio X, then after a year we have wealth equal to

$$W(1+r_X).$$

The expected utility of investing in X is therefore given by

$$\begin{aligned}\mathbb{E}(U(W(1+r_X))) &= \int_{-K}^{K} U(W(1+s))f_X(s)ds,\\ &= \int_{-K}^{K} U(W(1+s))\frac{d}{ds}F_X(s)ds.\end{aligned}$$

Since our hypotheses are on F_X and U', we can integrate by parts to move the derivative onto U. That is, we set $u(s) = U(W(1+s))$, and $v = F_X(s)$ in the integration by parts formula. Since

$$u'(s) = WU'(W(1+s)),$$

we obtain

$$\begin{aligned}&\int_{-K}^{K} U(W(1+s))\frac{d}{ds}F_X(s)ds\\ &\qquad = U(W(1+K)) - \int_{-K}^{K} WU'(W(1+s))F_X(s)ds.\end{aligned}$$

Repeating the same steps for Y and subtracting, we find that

$$\mathbb{E}(U(X)) - \mathbb{E}(U(Y)) = -\int_{-K}^{K} WU'(W(1+s))(F_X(s) - F_Y(s))ds. \qquad (11.1)$$

Since the derivative of U is positive and $F_X(s) - F_Y(s)$ is always non-positive and sometimes negative, we have that the difference in expected utilities is positive. Hence, as we set out to prove, X is preferred to Y. □

Suppose next we apply this result to the simplest increasing utility function:

$$U(W) = W.$$

If X first-order stochastically dominates Y, we have by our theorem that it will be preferred to Y and

$$\mathbb{E}(1 + r_X) = \mathbb{E}(U(1 + r_X)) > \mathbb{E}(U(1 + r_Y)) = \mathbb{E}(1 + r_Y). \tag{11.2}$$

We have shown that first-order stochastic dominance implies a greater expected return. Turning this round, we have that if two portfolios have the same expected return, first-order stochastic dominance **cannot** help us to distinguish them. To do so, we will need to adjust our assumptions.

11.4 Second-order stochastic dominance

The first-order stochastic dominance preference theorem works on the sole assumption that the investor prefers more to less. Importantly, it does not assume risk aversion and therefore cannot help us to take risk into account when choosing investments. This shortcoming is compounded by the fact that we are unlikely to find investments that satisfy the strong hypothesis of first-order stochastic dominance. An alternative might be to weaken the assumption on the returns at the cost of strengthening the assumption on utility to include risk aversion. With this in mind, we introduce the concept of second-order stochastic dominance using the integrals of the cumulative probability distributions.

Definition 11.6 We shall say that Y has second-order stochastic dominance over X if

$$\int_{-\infty}^{a} \mathbb{P}(r_X \leq s)ds \geq \int_{-\infty}^{a} \mathbb{P}(r_Y \leq s)ds \quad \text{for all } a \quad \text{and,}$$

$$\int_{-\infty}^{b} \mathbb{P}(r_X \leq s)ds > \int_{-\infty}^{b} \mathbb{P}(r_Y \leq s)ds \quad \text{for some } b.$$

We note that when dealing with discrete random variables, we can replace the integrals with finite summations. Provided the values are uniformly spaced, we simply have to compute the sums of the sums of the probability that each value is taken. If they are not uniformly spaced, we either have to use a finer subdivision to make them uniformly spaced, or to multiply the values by the distances between them; this reflects the fact that we are integrating step functions.

We might suspect that first-order stochastic dominance is a stronger condition than second-order stochastic dominance. We confirm this conjecture through the following reasoning. That is, suppose X is first-order stochastic dominant to Y. We wish to show that it is second-order stochastic dominant. To see this, however, it suffices to integrate the expressions

$$\begin{aligned} \mathbb{P}(r_X \le a) &\le \mathbb{P}(r_Y \le a) \quad \text{for all } a \quad \text{and,} \\ \mathbb{P}(r_X \le b) &< \mathbb{P}(r_Y \le b) \quad \text{for some } b. \end{aligned}$$

Hence we have that second-order stochastic dominance is indeed a weaker condition than first-order stochastic dominance. We are now in a position to state our second-order stochastic dominance theorem.

Theorem 11.7 *If portfolios X and Y have returns r_X and r_Y, and the investor has a utility function U with*

$$U'(s) > 0, \quad U''(s) < 0,$$

for all s, and the cumulative probabilities satisfy

$$\begin{aligned} \int_{-\infty}^{a} \mathbb{P}(r_X \le s)ds &\le \int_{-\infty}^{a} \mathbb{P}(r_Y \le s)ds \quad \textit{for all } a, \\ \int_{-\infty}^{b} \mathbb{P}(r_X \le s)ds &< \int_{-\infty}^{b} \mathbb{P}(r_Y \le s)ds \quad \textit{for some } b, \end{aligned}$$

then X will be preferred to Y.

Proof Our assumptions on U are that $U' > 0$, and $U'' < 0$. Our objective is to show that

$$\int U(W(1+s))f_X(s)ds > \int U(W(1+s))f_Y(s)ds$$

and hence X is preferred to Y. We proceed once more using integration by parts to re-express this inequality in the form

$$\begin{aligned} -\int_{-K}^{K} WU'(W(1+s))F_X(s)ds + U(W(1+K)) > \\ -\int_{-K}^{K} WU'(W(1+s))F_Y(s)ds + U(W(1+K)). \end{aligned}$$

Canceling terms and dividing by W, we have

$$-\int_{-K}^{K} U'(W(1+s))(F_X(s) - F_Y(s))ds > 0.$$

In order to obtain an expression involving U'', we integrate by parts again defining

$$\tilde{F}_X(s) = \int_{-\infty}^{s} F_X(t)dt, \tag{11.3}$$

$$\tilde{F}_Y(s) = \int_{-\infty}^{s} F_Y(t)dt, \text{ so} \tag{11.4}$$

$$F_X(s) = \frac{d}{ds}\tilde{F}_X(s), \tag{11.5}$$

$$F_Y(s) = \frac{d}{ds}\tilde{F}_Y(s). \tag{11.6}$$

We obtain

$$W\int_{-K}^{K} U''(W(1+s))(\tilde{F}_X(s) - \tilde{F}_Y(s))ds \\ - U'(W(1+K))(\tilde{F}_X(K) - \tilde{F}_Y(K)) > 0. \tag{11.7}$$

Since $U'' < 0$, $U' > 0$, this will be positive if

$$\tilde{F}_X(a) \leq \tilde{F}_Y(a) \quad \text{for all } a,$$
$$\tilde{F}_X(b) < \tilde{F}_Y(b) \quad \text{for some } b.$$

Since (11.7) must hold for all such U these will be necessary and sufficient conditions □

In summary, we have seen that the first-order stochastic dominance theorem makes assumptions about the cumulative distribution functions and requires the investor to prefer more to less. Second-order stochastic dominance makes weaker assumptions about the integral of the cumulative distribution functions but requires the investor to be risk averse as well as preferring more to less. First-order stochastic-dominance is certainly a stronger assumption to impose on the assets, but requires weaker assumptions on investors. We next illustrate second-order stochastic dominance with an example of a computational problem.

Example 11.8 Suppose we have two portfolios X and Y with returns distributed as shown in the table below. Our problem is to compare them using stochastic dominance.

X		Y	
return	probability	return	probability
5	0.1	5	0.1
6	0.3	6	0.1
8	0.1	7	0.1
9	0.2	8	0.3
12	0.3	10	0.1
		11	0.3

To tackle this problem, we build a table with rows as follows:

- a possible return r from either investment, in increasing order;
- its probability f for each investment;
- its cumulative probability F for each investment;
- the sum of the cumulative probabilities for each investment from higher rows $\tilde{F}$.

We also include $\log(1+r)$, and r^2 here since they are useful for computing other measures, resulting in a table as shown below.

r	log(1+r)	r^2	f_X	F_X	$\tilde{F}_X$	f_Y	F_Y	$\tilde{F}_Y$
2	0.020	4	0	0	0	0	0	0
3	0.030	9	0	0	0	0	0	0
4	0.039	16	0	0	0	0	0	0
5	0.049	25	0.1	0.1	0	0.1	0.1	0
6	0.058	36	0.3	0.4	0.1	0.1	0.2	0.1
7	0.068	49	0	0.4	0.5	0.1	0.3	0.3
8	0.077	64	0.1	0.5	0.9	0.3	0.6	0.6
9	0.086	81	0.2	0.7	1.4	0	0.6	1.2
10	0.095	100	0	0.7	2.1	0.1	0.7	1.8
11	0.104	121	0	0.7	2.8	0.3	1	2.5
12	0.113	144	0.3	1	3.5	0	1	3.5

We now use this to compute the differences of the cumulatives and their integrals:

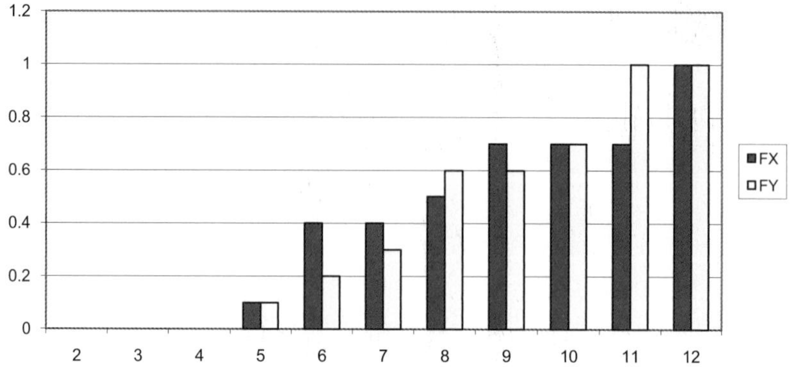

Figure 11.1 Cumulative distribution functions for the two portfolios X and Y.

$F_X - F_Y$	$\tilde{F}_X - \tilde{F}_Y$
0.00	0.00
0.00	0.00
0.00	0.00
0.00	0.00
0.20	0.00
0.10	0.20
-0.10	0.30
0.10	0.20
0.00	0.30
-0.30	0.30
0.00	0.00

Hence, it is clear that Y displays second-order stochastic dominance over X but neither has first-order dominance over the other. Referring back to our other measures, we can use the table to compute the following:

mean r_X	8.5	mean r_Y	8.5
variance r_X	6.85	variance r_Y	4.25
$\mathbb{E}(\log(1+r_X))$	0.08129	$\mathbb{E}(\log(1+r_Y))$	0.08140

It is clear that Y wins on geometric mean, and mean–variance analysis as well as second-order stochastic dominance. We illustrate with the graphs of the cumulative distribution functions and their integrals in Figure 11.1 and Figure 11.2. ◇

We recall that first-order dominance implied a higher mean and therefore

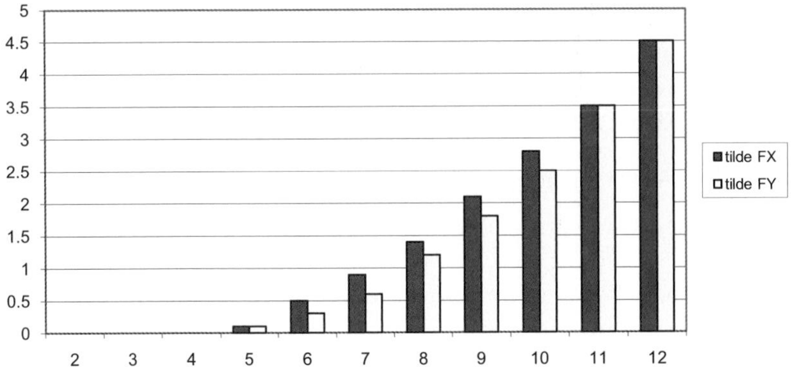

Figure 11.2 Integrals of cumulative distribution functions for X and Y

the technique is unable to distinguish between investments with the same mean return. Suppose now that we have portfolio X which is second-order stochastically dominant over Y. What can we say about their mean returns? We compute, again using integration by parts,

$$\begin{aligned}\mathbb{E}(r_X) &= \int_{-K}^{K} f_X(s)s\,ds, \\ &= F_X(K)K - \int_{-K}^{K} F_X(s)\,ds, \\ &= K - \tilde{F}_X(K).\end{aligned}$$

Since we assumed that $\tilde{F}_X(K) \leq \tilde{F}_Y(K)$, we have that

$$\mathbb{E}(r_X) \geq \mathbb{E}(r_Y).$$

Importantly, unlike the first-order case, second-order stochastic dominance may occur when the two portfolios share the same mean. Accordingly, we can use second-order stochastic dominance to distinguish between investments with the same mean. However, it will never tell us to prefer an investment with lower expected return on the basis that it is less risky. This is inevitable since we have made no assumptions as to how risk averse the investor is. They may have a tiny risk aversion or a huge one.

In a typical situation you may be asked to compare two investments X and Y using first- and second-order stochastic dominance. Possible answers will be

1. Nothing can be said.
2. X is first- and second-order stochastic dominant to Y. (Or Y to X.)
3. X is second-order stochastic dominant to Y but not first-order stochastic dominant. (Or Y to X.)

Example 11.9 Three assets X, Y and Z have returns distributed as follows:

- X : a continuous uniform on $[-10, 10]$;
- Y : takes values $-5, 0, 5$ with equal probabilities;
- Z : takes values $-10, 0, 10$ with equal probabilities.

Order these assets as far as possible for each of the following investors:

- an investor who uses an increasing utility function about which nothing is known;
- an investor who uses an increasing concave utility function for which nothing further is known.

Give brief justification.

The first investor can be assessed using first-order stochastic dominance. However, all three investments have the same mean which contradicts the existence of any first-order results. We cannot order at all.

For the second investor, we can use second-order stochastic dominance. First, if we compute $\mathbb{E}(\log(1+r))$, we get

$$5(\log(1.1)1.1 - 1.1 - \log(0.9)0.9 + 0.9) = -0.00167$$

for X and the other two give -0.000834377, and -0.003350112. Since the log utility function is a concave utility function, this means that any general results have to obey $Y > X > Z$.

The cumulative of X is 0 for below -10 and it is

$$(x+10)/20 = x/20 + 0.5,$$

from -10 to 10. Its integral is therefore

$$x^2/40 + 0.5x + 2.5.$$

The integral of the cumulative of Z grows linearly from -10 and X grows quadratically. In addition, Y is zero from -10 to -5. So the only possible relations are Y better than X, Z; and X better than Z.

At 10 all have cumulative integral equal to 10 since this is 10 minus the mean.

At 0, Z's cumulative integral is 3.33 which is bigger than X's. Given that the function for X is convex and the one for Z is linear on $(-10, 0)$ and $(0, 10)$, the one for X must be lower on $(-10, 10)$ and we conclude that X is second-order stochastic dominant to Z.

On $(-10, -5)$, Y is clearly below Z. At 0 they give $10/3$ and $5/3$. At 5 they give $10/3 + 10/3$ and $5/3 + 10/3$. At 10 they agree. Since the functions are piecewise linear, we have that Y is second-order stochastic dominant to Z.

That leaves X and Y. At -5, Y is below. At zero, X is $2.5 > 5/3$ so Y is lower. At 5, Y's cumulative integral is 5 and X's is $25/40 + 5$ so Y is lower.

At 10 they agree and X has slope 1 and Y also has slope 1. The function for Y is therefore the tangent to the convex function for X and lies below.

So Y is second-order stochastically dominant to X. We have

$$Y > X > Z.$$ ◇

To conclude this chapter, it is worth observing that the integration by parts technique used in our proofs of the first and second-order dominance theorems may be used iteratively to obtain higher-order dominance statements. Each time we integrate by parts, we get more conditions on higher-order derivatives of U together with conditions on iterated integrals of the cumulative distribution functions.

11.5 Review

By the end of this chapter, the reader should be able to do the following questions and tasks.

1. What does it mean for an investment to be dominant, first-order stochastically dominant and second-order stochastically dominant to another investment?
2. If X is dominant to Y, must it be more efficient in a mean–variance sense?
3. Under reasonable assumptions which should be stated clearly, prove that a portfolio that is first-order stochastically dominant to another investment will be preferred.
4. Does first-order stochastic dominance imply second-order stochastic dominance? Justify your answer. What about the other way round?
5. If X is first-order stochastically dominant to Y, what can we say about their expected returns? What about second-order? Justify your answers.

11.6 Problems

Question 11.1 Rank the following portfolios using first- and second-order stochastic dominance:

X		Y	
return	prob	return	prob
1	0.25	2	0.2
4	0.5	4	0.5
9	0.25	7	0.3

Question 11.2 Rank the following portfolios using first- and second-order stochastic dominance:

X return	X prob	Y return	Y prob
2	0.25	2	0.2
5	0.5	4	0.5
6	0.25	7	0.3

Question 11.3 Rank the following portfolios using first and second-order stochastic dominance:

X return	X prob	Y return	Y prob
2	0.1	2	0.2
5	0.4	4	0.5
7	0.5	7	0.3

Question 11.4 Consider the probabilities for rates of return on investments X, Y and Z below. Assuming iid returns, rank them as far as possible for an investor who is known to have a strictly increasing utility function. Repeat for one with an increasing concave utility function.

p	R_X
0.2	6
0.4	8
0.3	9
0.1	11

p	R_Y
0.1	7
0.3	8
0.2	9
0.3	10
0.1	11

p	R_Z
0.4	8
0.3	9
0.2	10
0.1	11

Question 11.5 Two investments A_1 and A_2 have returns such that the return of A_j is uniformly distributed on the interval $[a_j, b_j]$. Give necessary and sufficient conditions on the values of a_1, a_2, b_1 and b_2, so that:

- A_1 is preferred to A_2 for risk-neutral investors;
- A_1 is preferred to A_2 for all investors with increasing utility functions.

Justify your answer.

Question 11.6 Two investments A_1 and A_2 have returns such that the return of A_j takes values a_j and b_j with equal probability. Give necessary and sufficient conditions on the values of a_1, a_2, b_1 and b_2, so that:

- A_1 is preferred to A_2 for risk-neutral investors;
- A_1 is preferred to A_2 for all investors with increasing utility functions;

- A_1 is preferred to A_2 for all investors with increasing concave utility functions.

Justify your answer.

Question 11.7 Assets X, Y, and Z have the following yearly returns which depend on the state of the world which occurs:

State	X	Y	Z	Probability
A	5	6	5	0.4
B	7	4	5	0.3
C	–1	–2	5	0.2
D	11	19	5	0.1

Investors must place all of their money in precisely one of these assets. For each of the following investors, rank the assets as far as possible:

1. an investor who wishes to maximise very long term returns;
2. an investor who makes decisions according to an increasing linear utility function;
3. a mean–variance investor;
4. an investor with a log utility function;
5. an investor who makes decisions according to utility theory and prefers more to less;
6. an investor who makes decisions according to utility theory, prefers more to less and is risk averse.

12
Risk measures

12.1 Introduction

In our work so far we have largely used variance (or standard deviation) as the principal measure of risk, with occasional references to semi-variance. Whilst variance is a useful measure of risk, it has a number of shortcomings, including:

- it penalises upside variance as well as downside variance;
- mean–variance investors have quadratic utility functions, which since they are not increasing functions, are unrealistic;
- for general distributions which may be very different from normals, it can be too simplistic;
- for controlling trading books, management and regulators care about how much we stand to lose in the next ten days.

With these factors in mind, alternative measures of risk have become widespread in the finance industry, the most simple and popular of which has been "Value-at-Risk" or "VAR." As we will see, this is far from a perfect measure and it is not difficult to construct examples where VAR does not encapsulate risk particularly well. Better measures do exist and we will examine one, "conditional expected shortfall," in detail. We will then analyse the various risk measures we have introduced in terms of a new concept of "coherence," the intention of which is to encapsulate the ideal properties of a risk measure. We will then return briefly to utility theory and see which of our risk measures may be represented with a utility function. Finally, we discuss a major application of portfolio modelling in determine a bank's economic capital requirements.

12.2 Value-at-Risk

This is the most popular measure used for controlling trading risk in the finance industry and has enjoyed popularity with regulators for determining bank capital requirements to be held against certain risks. In essence, the idea is to determine how much an investor would stand to lose. The answer to this is, of course, everything, unless an investment is made in some riskless securities, if such truly exist. Refining this, we might ask how much an investor would stand to lose at a specified level of probability across a given time interval. The resulting VAR amount is therefore usually expressed in terms of dollar losses rather than returns. For example, how much could we lose with 5% probability in one day. Flipping this around, we would be 95% confident of our daily losses being contained within this level. This dollar amount becomes a number from which the board or management can more readily assess their appetite for risk. For example, if we were to set a 5% daily VAR limit of $2,000,000, then we understand that daily losses over $2,000,000 may happen, but they are relatively unlikely. If we were still uncomfortable with this, then we could consider reducing either the VAR limit at 5% or tightening the probability. For instance, when banks use VAR to determine how much capital they need to hold to absorb unexpected losses, the probability is set at a very low level, such as 0.03%. This is in recognition of the fact that were losses to exceed capital, the bank would have failed, and this would be a catastrophic fate. This is sometimes referred to as a "one in three-thousand year loss", since we can think of the VAR at probability level p as the amount one might expect to lose every $1/p$ years. The psychological effect of this viewpoint however may be to lull the unwary into a false sense of security with the idea that catastrophe is somehow too remote to occur in one's lifetime. Unexpected events can and do occur.

We now formalise our introduction to VAR with a proper definition.

Definition 12.1 Suppose that the value of a portfolio today is V_0 and at time t is V_t. We define the loss distribution to be the distribution of differences

$$L_t = V_0 - V_t,$$

where a negative value for L_t reflects an amount of profit.

The VAR at probability p for time period t is the value x such that

$$\mathbb{P}(L_t \geq x) = p.$$

We will always take VAR to be positive or zero. So if at probability p we make money, we will set the VAR to be zero.

Alternatively, if use F_t to denote the cumulative distribution function of L_t then the VAR at level p is

$$F_t^{-1}(1-p).$$

We note that in banking practice it is common to find VAR defined in terms of $1-p$, instead, for example, at the 95% level or 99% level.

There may not be a level x at which the probability of losing x or more is precisely p. This failure can only occur when the distribution is not continuous. However, there will instead be a level at which the probability jumps across p and we use that instead; we illustrate this in the next example.

Example 12.2 A portfolio A loses \$10,000,000 with probability 0.005, loses \$5,000,000 with probability 0.02, loses \$1,000,000 with probability 0.05. Otherwise, it makes \$1,000,000. Our problem is to find the VAR at 1 and 5 percent levels.

In accordance with our assumptions, we have

$$\begin{aligned}\mathbb{P}(L_t \geq 10) =&0.005,\\ \mathbb{P}(L_t \geq 5) =&0.025,\\ \mathbb{P}(L_t \geq 1) =&0.075 = 0.005 + 0.02 + 0.05.\end{aligned}$$

The 1% VAR level is thus \$5,000,000, since if $y > 5$,

$$\mathbb{P}(L_t > y) = 0.005 < 0.01,$$

and if $1 < z < 5$,

$$\mathbb{P}(L_t > z) = 0.025 > 0.01.$$

Similarly, the 5% VAR level is \$1,000,000. The cumulative probability distribution is a step function with 1% and 5% VAR levels shown in Figure 12.1.◇

One problem with VAR is that it does not provide any information about to what happens beyond level p. This may give a misleading impression as to the relative risks of two portfolios. We illustrate this in the following example where one has a risk of large loss albeit at a very low probability.

Example 12.3 Suppose we have an asset A which is worth 1 initially and at time t has value distributed as follows:

$$\begin{cases} 1.05 \text{ with probability } 0.9; \\ 0.9 \text{ with probability } 0.1. \end{cases}$$

Suppose further that the asset B is worth 1 initially and at time t has value

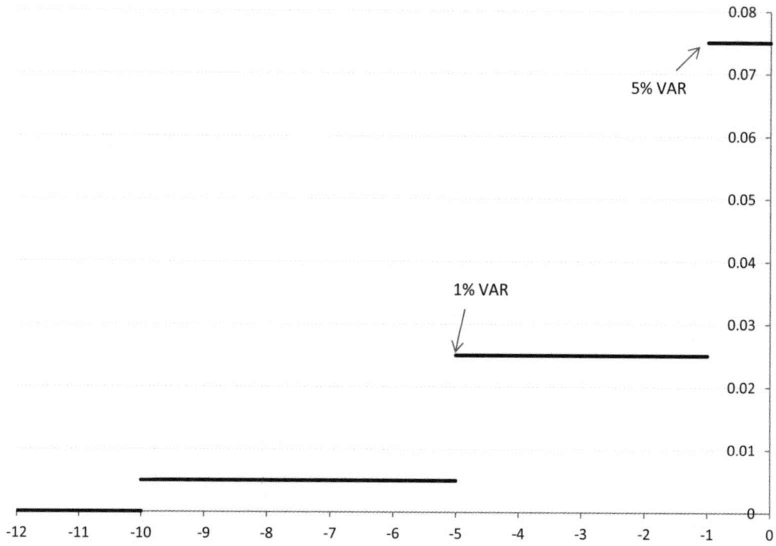

Figure 12.1 Cumulative probability distribution for the profit and loss of the portfolio in Example 12.2.

distributed as follows:

$$\begin{cases} 1.1 \text{ with probability } 0.99; \\ 0 \text{ with probability } 0.01. \end{cases}$$

If we work at a 5% VAR level, B appears to be low in risk since its VAR is negative which we take to be zero. However, at the same level A has a VAR of 0.1. At the 1% level however, the VAR of B jumps to 1.0, that is we have a small probability of losing everything! ◇

In practical finance this situation is not unusual. The probability of a creditworthy company failing to repay a loan is small, but when it does, the losses may be large: the probability may be so low that the risk does not show up at our VAR threshold. This aspect of VAR is therefore an issue for the practitioner and motivates the use of alternative measures, such as conditional expected shortfall which we examine later in this chapter.

12.3 Computing VAR

We need to be able to compute with VAR. Whilst for a general distribution, this can be very complex, is not difficult in the case where the loss distribution is normal. To see this, recall that if $X = N(0,1)$ is a standard normal distribution then the mean is 0, the variance is 1, and the density is

$$N'(x) = \frac{1}{\sqrt{2\pi}} e^{-\frac{x}{2}^2}.$$

Hence, we have that

$$\mathbb{P}(X \leq L) = N(L) = \frac{1}{\sqrt{2\pi}} \int_{-\infty}^{L} e^{-\frac{x}{2}^2}.$$

If Y is normally distributed with mean μ and variance σ^2 then we can write

$$Y = \mu + \sigma X.$$

The VAR of Y at level p is the number L such that $\mathbb{P}(Y \leq L) = p$, which is the number such that

$$\mathbb{P}(\sigma X + \mu \leq L) = p,$$

or alternatively

$$\mathbb{P}(X \leq \sigma^{-1}(L - \mu)) = p.$$

This means

$$N(\sigma^{-1}(L - \mu)) = p,$$

and hence,

$$\sigma^{-1}(L - \mu) = N^{-1}(p).$$

Therefore, we conclude

$$L = \mu + \sigma N^{-1}(p). \tag{12.1}$$

Readers familiar with EXCEL will observe that we can compute N^{-1} using the NORMSINV() function. Some typical values are shown in Table 12.1.

We note however that since this formula for VAR only involves the mean and variance of the portfolio and a normal distribution is fully determined by its mean and variance, we achieve no further insights from VAR.

Example 12.4 If a portfolio has expected value 100 and standard deviation 1000, what is the VAR at a 1% level? The answer is

$$100 + 1000N^{-1}(0.01) = -2226.35.$$

p	$N^{-1}(p)$
0.001	–3.090232306
0.01	–2.326347874
0.05	–1.644853627
0.2	–0.841621234

Table 12.1 *Inverse cumulative normal probabilities.*

Since VAR is usually quoted in terms of losses, the VAR is 2226.35. 

Furthermore, we note that given a collection of assets, A_i, with jointly normal distribution and covariance matrix C, then a portfolio with weights X_i will also be normally distributed, and we can compute its VAR from its mean and variance using the above method.

We conclude this discussion with a word of caution: whilst this method is simple to compute, real-world financial losses tend not to be normally distributed. Indeed these distributions tend to have "fat tails," where the probability of being far from the mean is greater than that for a normal distribution with the same mean and variance. This can often be summarised by looking at the *kurtosis* or fourth moment:

$$\frac{\mathbb{E}((X-\mu)^4)}{\text{Var}(X)^2}.$$

For a normal distribution, this is equal to 3. For fat-tailed distributions it will be higher. Applying a normal approximation to a fat-tailed distribution will lead to VAR numbers that are too low, since the probability of large moves is underestimated.

In practice, computation of VAR numbers often involves the use of Monte Carlo simulation models. By running a very large number of paths and calculating the loss outcome in each instance, the required loss percentile may be estimated. The integrity of the output depends on the assumptions, including those concerning correlations, made in modelling the underlying random elements. For example, for a portfolio of bank loans, the loss distribution will depend on both the individual loss distributions of the underlying loans and the correlation structure between these distributions. What is the correlation of default between a residential construction company and an airline? Here the bank portfolio manager is confronted with the very real limitations of historic data, yet in calculating a bank's "economic capital" this is precisely what he or she needs to do.

12.4 VAR estimates and excesses

Suppose that we are market risk managers responsible for monitoring a trading portfolio with 5% daily VAR (i.e. for the time period of a day) for 220 trading days. The accuracy of our VAR calculations will depend on the accuracy of the loss distribution which we are using. Since this is a modelled distribution, we need to assess whether outcomes are in line with the modelled VAR. Certainly, if the VAR is correct, we would expect the daily losses to exceed the VAR level on a number of occasions according to the risk threshold we have set, in this case 5%. A day on which the loss exceeds the VAR level is called an excess. The expected number of such events across the 220 trading days is simply

$$220 \times 0.05 = 11.$$

It follows that if there were many more excesses we should be worried. Likewise, if there were far fewer, our estimate of VAR is probably too conservative and we should also be concerned. Whilst effective at limiting risk, an overly conservative measure will inhibit new business and reduce our opportunity for profit. The process of looking back over a period to see whether the number of excesses that occurred was reasonable is sometimes called *back-testing*.

Generalising from the above example, if we monitor at level p for N periods, we can compute the probability distribution of excesses. It will be binomial with probability p of an excess event in each period. Hence in the above example, the probability of zero excesses for the 220-day period is

$$(1-0.05)^{220}.$$

12.5 Evaluating risk measures

Having introduced VAR, and remarked upon its popularity for managing trading risks and bank capital models, we now step back and assess its properties as a risk measure. Does it behave as a good and sensible risk measure ought? To address this question, we need to decide precisely what we mean by this and start by defining a risk measure to be a map from the set distributions of possible *changes* in a portfolio's value across a fixed time period to the real numbers. The fixed time period, sometimes referred to as the "risk horizon" will depend upon the financial risks that we are seeking to measure. Trading risks tend to be measured daily, whereas investment risks may be measured across a horizon of one or more years.

Definition 12.5 Denote by Δ the set of possible probability distributions of changes in value. A risk measure is a map

$$\rho : \Delta \to \mathbb{R}. \tag{12.2}$$

We emphasise that an element of Δ will be a distribution of changes in value. For a given portfolio V and time horizon t, we represent the associated distribution of changes by X_t^V. By convention, we will reflect a risk of loss with a positive value of ρ. For instance, if a portfolio has a 5% risk of losing \$20,000,000, then its VAR at the 5% level is \$20 million rather than $-\$20,000,000$.

The first important property which we would like our risk measures to satisfy is *monotonicity*. That is, if two portfolios V and W are of the same initial value yet V always returns more than W, then we conclude that V carries less risk than W. We would want our risk measure to reflect this and hence the risk measure, ρ, is said to be monotone if in this case, we always have

$$\rho(X_t^V) < \rho(X_t^W).$$

It is not difficult to show that VAR is a monotone risk measure. To see this, suppose we have portfolios V and W for which

$$W_0 = V_0$$

and

$$W_t \geq V_t.$$

Then

$$-V_t \geq -W_t.$$

Let L_t^V be the losses for V and L_t^W for W. We then have

$$L_t^V = V_0 - V_t \geq W_0 - W_t = L_t^W.$$

Hence if

$$\mathbb{P}(L_t^W \geq x) = p,$$

then

$$\mathbb{P}(L_t^V \geq x) \geq p,$$

and the VAR for V is at least equal to the VAR for W.

A second intuitive property of a risk measure is that the sum of the risks of two portfolios considered separately should be more than or equal to that of two portfolios considered together. That is, the larger portfolio has the opportunity for diversification and hence reduction of risk. Hence we say that a risk measure ρ is *subadditive* if for two portfolios V and W and time horizon t

$$\rho(X_t^V + X_t^W) \leq \rho(X_t^V) + \rho(X_t^W).$$

We note that we do not require equality since the portfolios V and W could be natural hedges with a consequent negative correlation. An extreme case would be if $W = -V$, in which case $W + V$ would be riskless.

Interestingly, it is subadditivity which brings to our attention a shortcoming of VAR. To demonstrate the failure of subadditivity for VAR, it suffices to construct an example. We do this now.

Example 12.6 Suppose assets C and D are independent and worth 1 initially. Each is worth

$$\begin{cases} 1.1 \text{ with probability } 0.96, \\ 0.5 \text{ with probability } 0.04. \end{cases}$$

What is the VAR at 5% for each individually and for both together? The VAR of each individual asset at 5% is zero. We will make no loss with 95% probability. The initial value of the portfolio is 2.0 and the joint distribution of the final values will be as follows:

$$\begin{cases} 2.2 \text{ with probability } 0.96^2 = 0.9216, \\ 1.6 \text{ with probability } 2 \times 0.96 \times 0.04 = 0.0768, \\ 1.0 \text{ with probability } 0.04^2 = 0.0016. \end{cases}$$

The first case represents a profit of 0.2, whereas the second and third cases reflect losses of 0.4 and 1.0 respectively. The 5% VAR will be 0.4 since the probability of having a loss greater than 0.4 is less than 5% and for any $0 < z < 0.4$ the probability of a loss greater than or equal to z is greater than 5%. Since the VAR of the combined portfolio at 5% is higher than the sum of the individual VARs, we have shown that VAR fails subadditivity. ◇

Returning to our search for those characteristics we would like a risk measure to have, our third property is that a risk measure should be scaleable. That is, if we multiply everything in our portfolio by the same amount (and hence the associated distribution of value changes) then the risk should grow by the same factor.

Definition 12.7 A risk measure ρ is said to be *positively homogeneous* if for $X_t^V \in \Delta$, $h > 0$,

$$\rho(hX_t^V) = h\rho(X_t^V).$$

Note that we only require this for $h > 0$. Holding a negative amount of an asset should not change the sign of the risk measure.

It is not difficult to show that VAR satisfies positive homogeneity. That is, suppose that we compute the VAR at probability level p for a portfolio V over time horizon t to be an amount L and let $L_t^V = V_0 - V_t$ denote the distribution of losses for V. For $h > 0$, we have that

$$\mathbb{P}(hL_t^V \leq hL) = \mathbb{P}(L_t^V \leq L) = p.$$

So

$$\mathrm{VAR}(hL_t^V) = h\,\mathrm{VAR}(L_t^V).$$

Our final property considers the case where we add or subtract a fixed sum of money from the changes of a portfolio. The risk measure should change by the same amount.

Definition 12.8 A risk measure ρ is said to be *translation invariant* if $X_t^V \in \Delta$, $a \in \mathbb{R}$, then

$$\rho(a + X_t^V) = \rho(X_t^V) - a.$$

To see that VAR is translation invariant, we note that adding a fixed sum of money to a portfolio moves the entire distribution, and therefore each loss percentile, by the same amount. That is,

$$\mathbb{P}(L_t^V \leq L) = p \implies \mathbb{P}(L_t^V + a \leq L + a) = p.$$

So

$$\mathrm{VAR}(L_t^V + a) = -(L + a) = \mathrm{VAR}(L_t^V) - a.$$

Having mathematically defined the properties we would like to see in a well-behaved risk measure, we may use these properties as axioms.

Definition 12.9 If a risk measure satisfies the four axioms of monotonicity, subadditivity, positive homogeneity and translation invariance, then it is said to be a *coherent* risk measure.

We have seen that VAR is not coherent although it satisfies all of the axioms except subadditivity. In addition, VAR can jump whilst the portfolio is changing in a continuous fashion. We would not expect a small change in a portfolio to result in a significant change in its risk. However, when using VAR as the risk measure this *can* happen, as shown in the below example.

Example 12.10 Suppose asset X_p loses $1,000,000$ with probability p and zero otherwise. Then at the 5% level, we have

$$\mathrm{VAR}(X_p) = \begin{cases} 1,000,000 \text{ for } p \geq 5\%, \\ 0 \text{ for } p < 5\%. \end{cases}$$

A small change in p only changes the asset slightly. The consequent VAR change may be huge. In other words VAR is not continuous. ◇

Despite these shortcomings, VAR still remains the dominant risk measure in trading environments. This is partly driven by the regulators and partly a consequence of historical accident: VAR (after variance) was the second measure to be thought of. Later in this chapter we examine a potential improvement to VAR which satisfies not only the four axioms of coherence, but continuity as well. Before then, however, we examine our more familiar risk measures, variance, standard deviation and semi-variance to see how they perform in terms of the coherence axioms.

12.6 Other risk measures and the axioms

We have seen that VAR fails subadditivity but satisfies the other three coherence axioms. How well then do our other more familiar risk measures perform? We start by investigating variance.

Firstly, we observe that variance is not monotone by constructing a simple counterexample. That is, suppose we have a profit and loss distribution X such that $X = 1$ with probability 0.5 and $X = 2$ with probability 0.5. Suppose furthermore that we have a second profit and loss distribution for which $Y = 0.5$ with probability 0.5 and $Y = 0.75$ with probability 0.5. Clearly, X always returns more than Y but the variance of X is 16 times as large as that of Y (i.e. 0.25 compared with 0.015625.)

In relation to subadditivity, recall that

$$\mathrm{Var}(X+Y) = \mathrm{Var}(X) + \mathrm{Var}(Y) + 2\,\mathrm{Cov}(X,Y).$$

Hence in the event that the covariance is positive, the sum of the variances is less than the variance of the sum. We conclude that variance is not subadditive.

For positive homogeneity, we note that for any h, we have

$$\mathrm{Var}(hX) = h^2\,\mathrm{Var}(X),$$

and hence variance is not positively homogeneous.

Finally for translation invariance we observe that variance involves subtracting the mean. It follows that adding a positive constant will not affect the variance, that is,

$$\mathrm{Var}(X+a) = \mathrm{Var}(X).$$

Hence translation invariance fails as well.

We conclude that variance, our oldest and in many ways most fundamental risk measure has failed all the axioms of coherence. We next examine standard deviation to see if this performs any better in this regard. Firstly, we have that standard deviation fails monotonicity for the same reasons as variance and the same counterexample suffices. For subadditivity we have that

$$\mathrm{Var}(X+Y) = \mathrm{Var}(X) + \mathrm{Var}(Y) + 2\,\mathrm{Cov}(X,Y),$$

but

$$\mathrm{Cov}(X,Y) = \sigma_X \sigma_Y \rho_{XY} \leq \sigma_X \sigma_Y.$$

So

$$\sigma_{X+Y}^2 \leq \sigma_X^2 + \sigma_Y^2 + 2\sigma_X\sigma_Y = (\sigma_X + \sigma_Y)^2.$$

Taking square-roots we have that standard deviation is subadditive.

Furthermore $\mathrm{Var}(hX) = h^2\,\mathrm{Var}(X)$, and hence taking square roots we get for $h > 0$,

$$\sigma_{hX} = h\sigma_X.$$

Therefore standard deviation is positively homogeneous.

For translation invariance, we have already seen that $\mathrm{Var}(X+a) = \mathrm{Var}(X)$, and thus standard deviation fails translation invariance for the same reason. We therefore have that standard deviation passes two of the four coherence axioms.

Finally, we recall that semi-variance is defined to be

$$S(W) = \mathbb{E}((W - \mathbb{E}(W))^2 H(\mathbb{E}(W) - W))).$$

where H is defined such that

$$H(s) = \begin{cases} 1 \text{ for } s \geq 0, \\ 0 \text{ for } s < 0. \end{cases}$$

The semi-variance is not monotone for similar reasons to variance. Indeed, the counterexample we constructed for variance yields a similar result for semi-variance. Again, we find that for subadditivity and homogeneity, semi-variance will fail for the same reasons as variance. For example, suppose $W = 2$ or -2 with probability 0.5 each. We have that $\mathbb{E}(W) = 0$, and

$$S(W) = \mathbb{E}(W^2 H(-w)) = 0.5 \times 2^2 = 2;$$

however,

$$S(W+W) = S(2W) = 4\mathbb{E}(W^2 H(-W)) = 8.$$

Since semi-variance involves subtracting the mean, we have for any X that

$$S(X+a) = S(X).$$

Hence S is not translation invariant and (like variance) has failed all four coherence axioms.

12.7 Conditional expected shortfall

In order to overcome some of the problems we have seen with VAR, we introduce the concept of *conditional expected shortfall.* Instead of looking at the percentile loss level in isolation, this measure, which we denote by $\mathrm{CES}(X,p)$, considers the expected losses in the event the VAR level is exceeded. That is, for a profit and loss distribution X, across some fixed time horizon, we define

$$\mathrm{CES}(X,p) = -\frac{1}{p}\mathbb{E}(XH(X_p - X))$$

and X_p is such that $\mathbb{P}(X \leq X_p) = p$. Assuming X_p is a negative quantity, then X_p is the negative of the level p VAR for X. In this case $H(X_p - X)$ will be zero for all values of X other than those negative amounts (losses) which lie in the tail of the distribution beyond X_p. Furthermore since $\mathbb{E}(XH(X_p - X)$ will be a negative quantity, $\mathrm{CES}(X,p)$ will be positive. Equivalently, if X is a continuous distribution with density f we have

$$\mathrm{CES}(X,p) = -\frac{1}{p}\int_{-\infty}^{x} sf(s)ds \quad \text{where} \int_{-\infty}^{x} f(s)ds = p.$$

Example 12.11 Suppose the profit and loss distribution is normal with mean 0 and variance 1. If $N(L) = p$, we have

$$\begin{aligned}\mathrm{CES}(X,p) &= -\frac{1}{p}\frac{1}{\sqrt{2\pi}}\int_{-\infty}^{L} xe^{-x^2/2}dx, \\ &= \frac{1}{p}\frac{1}{\sqrt{2\pi}}\left[e^{-s^2/2}\right]_{-\infty}^{L}, \\ &= \frac{1}{p}\frac{1}{\sqrt{2\pi}}e^{-L^2/2}. \end{aligned} \tag{12.3}$$

A standard normal is, however, not very useful. If we have $Y = \mu + \sigma X$, then the CES will shift by μ and scale by σ :

$$\mathrm{CES}(Y,p) = \mu + \sigma\,\mathrm{CES}(X,p) = \mu + \frac{1}{p}\frac{\sigma}{\sqrt{2\pi}}e^{-L^2/2}, \tag{12.4}$$

where $L = N^{-1}(p)$. ◇

As well as computing CES by normal approximation, we can use the historical method. That is, suppose there are N observations. We take the maximal k such that $k/N \leq p$, and then calculate the average of the k worst observations.

Despite the prevalence of VAR, CES has some notable advantages. Most importantly, it is able to reflect information from beyond the VAR level. For example, if we are considering losses at a level of 5%, VAR will not change if the probability of a large unexpected loss increases so long as it stays below the fixed VAR level. By calculating the mean of these events, CES will increase accordingly. It follows that when a portfolio is lumpy – that is, it has some very big exposures with a small probability of big losses for each – CES comes into its own. Hence it is typically used when modelling portfolios of large loans or holdings of bonds. We illustrate with a computational example and compare the outcomes for CES and VAR.

Example 12.12 Suppose we have a portfolio comprising ten one-year loans with the following attributes:

- loan defaults are independent;
- each of 1,000,000 principal; and
- each loan has 0.4% chance of defaulting with no recovery value.

What is the VAR and CES at a 5% level?

We first calculate the probabilities of different default outcomes.

k	Probability No. defaults $\leq k$	Probability No. defaults $\geq k$	Probability No. defaults $= k$	k times probability
0	0.960712374	1	0.960712374	0
1	0.9992952	0.039287626	0.038582826	0.038582826
2	0.99999248	0.0007048	0.00069728	0.00139456
3	0.999999947	7.52026E-06	7.46752E-06	2.24026E-05
4	1	5.27364E-08	5.24826E-08	2.0993E-07
5	1	2.53777E-10	2.52928E-10	1.26464E-09
6	1	8.48432E-13	8.4648E-13	5.07888E-12
7	1	1.9984E-15	1.94258E-15	1.35981E-14
8	1	0	2.92557E-18	2.34046E-17
9	1	0	2.61095E-21	2.34986E-20
10	1	0	1.04858E-24	1.04858E-23

We conclude that the 5% VAR is 0, since the probability of any defaults occurring at all is less than 5%. The CES however is obtained by summing the terms in the right-most column, multiplying by the 1,000,000 loss, scaling, and dividing by 5%.

If we continue to compute the VAR at the 5% level for varying numbers of loans, we find that it is zero for any less than 13 loans and has therefore totally failed to detect the risk. We also see a big jump from 12 to 13 from zero to 1,000,000:

Loans	VAR	CES
10	0	800000
11	0	880000
12	0	960000
13	1000000	1040000

We see that CES is much better behaved and increases incrementally. ◇

A simplification of CES uses shortfall below a given fixed level. That is, suppose we have a profit and loss distribution X and we fix a threshold level L (a negative amount); then we can define expected shortfall $ES(X,L)$, to be

$$ES(X,L) = -\mathbb{E}(XH(L-X)).$$

Expected shortfall has the virtue of simplicity without penalising upside risk. Its chief virtue is that it is easy to explain.

12.8 CES and the coherence axioms

We will now show that, unlike VAR, CES satisfies all four of the axioms and hence is a coherent risk measure. To simplify our proofs, we will assume that the distribution of profit and loss X is continuous. Recall first that

$$\mathrm{CES}(X,p) = -\frac{1}{p}\int_{-\infty}^{X_p} sf(s)ds$$

where X_p is such that

$$\mathbb{P}(X \leq X_p) = p.$$

If $F(x)$ is the cumulative distribution function, we note that by substitution of variables, if we set $s = F(x)$ and $ds = f(x)dx$, then

$$\int_{-\infty}^{X_p} xf(x)dx = \int_{-\infty}^{p} F^{-1}(s)ds$$

and hence

$$\text{CES}(X,p) = -\frac{1}{p}\int_{-\infty}^{p} F^{-1}(x)dx.$$

In order to show monotonicity, suppose that we have two portfolios V_X and V_Y with profit and loss distributions X and Y respectively such that V_X always returns more than V_Y. We denote the cumulative profit and loss distribution functions by F_X and F_Y. Since V_X always returns more than V_Y it follows that

$$F_X(s) < F_Y(s)$$

for all values of s. Since the cumulative distribution function is a monotone increasing function we have that

$$F_Y^{-1}(x) < F_X^{-1}(x)$$

for any $x \in (0,1)$ and hence that

$$-\frac{1}{p}\int_{-\infty}^{p} F_Y^{-1}(x)dx > -\frac{1}{p}\int_{-\infty}^{p} F_x^{-1}(x)dx.$$

Hence $\text{CES}(Y,p) > \text{CES}(X,p)$. That is, the lower risk portfolio V_X has lower CES and we conclude that CES satisfies monotonicity.

We next want to show that CES is subadditive. That is suppose we have profit and loss distributions X, Y and Z such that $Z = X+Y$; we want to show

$$\text{CES}(Z,p) \leq \text{CES}(X,p) + \text{CES}(Y,p).$$

Let X_p, Y_p, Z_p be the negatives of the VAR measures at level p for each of X, Y and Z respectively. It suffices to show (noting minus signs) that

$$\mathbb{E}(ZH(Z_p - Z) - XH(X_p - X) - YH(Y_p - Y)) \geq 0. \tag{12.5}$$

We can rewrite this as

$$\mathbb{E}(X(H(Z_p - Z) - H(X_p - X))) + \mathbb{E}(Y(H(Z_p - Z) - H(Y_p - Y))) \geq 0. \tag{12.6}$$

We will show that each of these terms is non-negative. We examine the first term only, since the second will follow by symmetry. We can divide our argument into four cases according to the signs of the terms as follows:

$$\begin{aligned}
Z < Z_p, X < X_p, &\implies X(1-1) = 0 \geq X_p(1-1), \\
Z > Z_p, X > X_p, &\implies X(0-0) = 0 \geq X_p(0-0), \\
Z > Z_p, X < X_p, &\implies X(0-1) \geq X_p(0-1), \\
Z < Z_p, X > X_p, &\implies X(1-0) \geq X_p(1-0).
\end{aligned}$$

The term is therefore always greater than

$$X_p(H(Z_p - Z) - H(X_p - X)).$$

We have

$$\mathbb{E}(X_p(H(Z_p - Z) - H(X_p - X))) = X_p(\mathbb{E}(H(Z_p - Z)) - \mathbb{E}(H(X_p - X))),$$

but $\mathbb{E}(H(Z_p - Z)) = \mathbb{P}(Z \leq Z_p) = p$, and similarly for the other term so the whole thing is greater than or equal to $X_p(p - p) = 0$, and we are done. We conclude that subadditivity holds.

For positive homogeneity, we have

$$\text{CES}(\lambda X, p) = -\frac{1}{p}\mathbb{E}\big(\lambda X H((\lambda X)_p - \lambda X)\big).$$

But

$$(\lambda X)_p = -\text{VAR}(\lambda X) = \lambda X_p.$$

So since $\lambda > 0$, we have

$$\text{CES}(\lambda X, p) = -\frac{1}{p}\lambda\mathbb{E}(XH(X_p - X))) = \lambda\,\text{CES}(X, p),$$

and CES is positively homogeneous.

It remains to check translation invariance. We have that

$$\text{CES}(X + a, p) = -\frac{1}{p}\mathbb{E}((X + a)H((X + a)_p - (X + a))),$$

but

$$(X + a)_p = X_p + a,$$

and hence

$$\text{CES}(X + a, p) = \text{CES}(X, p) - \frac{1}{p}a\mathbb{E}((H(X_p - X))) = \text{CES}(X, p) - a.$$

In conclusion, we have shown that CES (unlike VAR) is a coherent risk measure.

An additional point, to be aware of, is that we showed that the theoretical value of CES is coherent; we did not show that approximations to CES are coherent. If we actually implemented CES we would be using an approximation, so ideally we should show that the approximation is coherent rather than the theoretical value.

12.9 Risk measures and utility

The use of risk measures such as variance, semi-variance, and shortfall can be related to properties of utility functions. We have already seen that mean–variance analysis is equivalent to working with a quadratic utility function. We now show that semi-variance corresponds to a utility function that is quadratic below the cut-off (either zero or the expected return) and linear above it. To see this recall the definition of semi-variance:

$$S(W) = \mathbb{E}((W - \mathbb{E}(W))^2 H(\mathbb{E}(W) - W)).$$

Suppose then that our utility is quadratic below the mean and linear above it. We obtain

$$U(W) = \begin{cases} a + b(W - \mathbb{E}(W)) & \text{for } W > \mathbb{E}(W), \\ a + b(W - \mathbb{E}(W)) - c(W - \mathbb{E}(W))^2 & \text{otherwise,} \end{cases}$$

for some a, b, c with $b, c > 0$. We can pass to an equivalent utility function such that $a = 0$, and $b = 1$ by taking a linear multiple of U. Hence our utility function is equivalent to one of the form

$$U(W) = \begin{cases} (W - \mathbb{E}(W)) & \text{for } W > \mathbb{E}(W), \\ (W - \mathbb{E}(W)) - c(W - \mathbb{E}(W))^2 & \text{otherwise.} \end{cases}$$

We can therefore write

$$U(W) = W - \mathbb{E}(W) - c(W - \mathbb{E}(W))^2 H(\mathbb{E}(W) - W).$$

We thus have

$$\mathbb{E}(U(W)) = -cS(W).$$

VAR and CES do not correspond naturally to utility functions, since they are defined too much in terms of percentiles.

12.10 Economic capital modelling

We conclude our chapter on risk measures with a brief discussion of economic capital modelling, an area in which many bank portfolio managers work. This arises from the regulatory requirement that banks operate their own internal portfolio model to compute how much capital they should hold. This capital, made up primarily of issued share capital, should be sufficient to absorb losses which may occur as a result of a bank's business activities to a very high level of confidence. Indeed, as noted earlier in this chapter, $p = 0.03\%$ is the most commonly used level. Firstly, the portfolio manager must be sure to have accounted for all risks with which the institution is confronted in the course of its daily business. The most obvious is credit risk, which is the risk that customers fail to repay their loans. This, however, is not the only risk for a modern institution. Interest rate risk, which might arise either through trading financial contracts or through the making of fixed rate loans funded by floating rate borrowings is another obvious source of risk to be modelled. Furthermore, there is operational risk, which includes losses arises from systems faults, natural disasters, frauds and, of course, simple human error. The bank portfolio manager has the task of building risk models for each of these risks, usually taking into whatever historic loss and correlation data is available and projections as to how these quantities might be impacted by external factors such as the economic cycle. For operational risk, this is very challenging. One approach is for risk managers to determine possible loss scenarios and to try to assign probabilities to these events. From this a loss distribution may be built using Monte Carlo simulation and the appropriate loss percentile used to determine operational risk capital.

Economic capital modelling is likely to be a focus of development for some years, the objective being to build a consistent framework which boards of directors and senior management may use to set and communicate to their stakeholders, for example shareholders, creditors and regulators, their risk appetite (or tolerances) for the different business activities of the bank. Measures such as "return on risk capital" which are often used to drive performance only become meaningful when all sources of risk are allocated a capital value.

12.11 Review

By the end of the chapter, the reader should be able to answer the following theoretical questions.

1. What does "VAR" stand for?
2. What is value at risk?
3. What does it mean for a risk measure to be subadditive? Prove or disprove that each of VAR and variance is subadditive.

4. What does it mean for a risk measure to be monotone? Prove or disprove that each of VAR and variance is monotone.
5. What does it mean for a distribution to be fat-tailed? How will the VAR of such a distribution compare to that of a normal distribution?
6. What is a VAR excess? What form does the distribution of the number of excess over a fixed period of time take?
7. If we change the size of a loss below the VAR level, what effect will it have on the VAR?
8. For each of the following risk measures, discuss how they relate to utility functions: shortfall; semi-variance; VAR; CES.

12.12 Problems

Question 12.1 A bank monitors its daily VAR at a 1% level. If the VAR is correct, how many trading days would it take for the probability of a VAR excess to be at least

- 50%,
- 99% ?

What is the probability of getting at least one excess in a ten-day period?

Question 12.2 Suppose that a trading book is monitored with a 10% one-day VAR limit. What is the probability of there being zero days where the VAR level was exceeded in a period of 10 trading days?

Question 12.3 Suppose that a trading book is monitored with a 5% one-day VAR limit. What is the probability of there being 0 or 1 days where the VAR level is exceeded in a period of 90 trading days?

Question 12.4 Suppose that a portfolio is managed with a 5% daily VAR. What is the probability of 0 excesses over 50 days and over 100 days? What is the probability of more than two excesses over the same time periods?

Question 12.5 Suppose that a portfolio is managed with a 1% daily VAR. What is the probability of 0 excesses over 50 days and 100 days. What is the probability of more than two excesses over the same time periods?

Question 12.6 A portfolio has value $\$100,000,000$. Its mean and standard deviation at time 1 are $\$110,000,000$ and $\$10,000,000$. Assuming it is normally distributed, what are its VAR at levels 1% and 5% ?

Question 12.7 Over a 90-day period, the trading losses of a bank exceed the one-day 1% VAR level three times. Would you regard this as a cause for concern? Justify your answer. What if it exceeded the VAR level two times or four times?

Question 12.8 A portfolio is worth \$10 billion. It is estimated to have the following distribution of value after a ten-day period:

- quarters in value with probability 0.0001;
- loses \$150,000,000 with probability 0.001;
- otherwise it is worth a uniformly distributed amount on the interval \$900 million to \$1.2 billion.

Find the 0.1%, 1% and 5% ten-day VARs. Suppose it is reestimated that instead of quartering in value in the worst case, it will half its value. What will the new VARs be?

12.13 Additional problems

Question 12.9 Suppose that profit and loss values are normally distributed across time with mean at time t of μt and standard deviation at time t of $\sigma\sqrt{t}$, with t measured in years. Compute one-day, one-year and five-year 5% VAR if $\mu = 0.2$ and $\sigma = 0.2$.

Now suppose that our estimates are only accurate to within 10% (in a relative sense). Compute bounds on the VAR varying each of μ and σ individually and then both together. Interpret your results.

Question 12.10 Suppose that each day a portfolio's value changes by a normal random variable X_j. These random variables have zero mean and are independent and identically distributed. How will ten-day VAR compare to one-day VAR?

13
The Capital Asset Pricing Model

13.1 Introduction

We recall from Chapter 4 that under certain assumptions, principally that an investor only cares about mean and variance, we are able to derive expression for a portfolio such that the investor would invest in a multiple of it and the risk-free asset. This portfolio was called the tangent portfolio. The tangent portfolio depends on various things that vary from investor to investor. If one makes assumptions that stops this variation, then one can develop a simple relationship between the expected return of an asset and its covariance with the market portfolio. This is called the *Capital Asset Pricing Model* or *CAPM*.

13.2 From tangent to market

We start by revisiting the algorithm developed in Chapter 4 for finding weights, x, of the tangent portfolio for a given risk-free rate R_f. That is, to find x, we solve the system

$$Cy = \overline{R} - R_f e, \qquad \text{and put}$$
$$x = \frac{y}{\langle y, e\rangle},$$

where we recall that

$$\overline{R} = (\overline{R}_1, \ldots, \overline{R}_n).$$

Our argument will comprise three parts, according to the following.

- In the first part all we assume is that a portfolio T has weights given by the algorithm above, and use that to deduce an equation relating a general portfolio's expected return to its covariance with T.

- In the second part, we assume that all investors have the same tangent portfolio, and use this to argue that T is, in fact, the market portfolio.
- In the third part, we examine the assumptions that would imply all investors have the same tangent portfolio.

We proceed with the first step by writing $y = \gamma x$ for some $\gamma \in \mathbb{R}$ where, of course, $\gamma = \langle y, e \rangle$. We then have

$$\gamma C x = \overline{R} - R_f e,$$

as a vector equation. Now suppose we take coordinate i, this then becomes

$$\gamma (Cx)_i = \mathbb{E}(R_i) - R_f,$$

which is equivalent to

$$\mathbb{E}(R_i) = R_f + \gamma (Cx)_i.$$

This equation relates the expected return on security i to the covariance matrix. However, the weights x are in fact the weights of the tangent portfolio and hence $(Cx)_i$ is the covariance between the tangent portfolio and asset i. We therefore conclude

$$\mathbb{E}(R_i) = R_f + \gamma \mathrm{Cov}(R_i, R_T).$$

Since γ is independent of i, the covariance of an asset with the tangent portfolio determines its expected return.

For this equation to be useful, we need to compute γ. We have

$$\begin{aligned}
\mathbb{E}(R_T) &= \sum_{i=1}^{n} x_i \mathbb{E}(R_i), \\
&= \left(\sum_{i=1}^{n} x_i R_f \right) + \gamma \sum_{i=1}^{n} x_i \mathrm{Cov}(R_i, R_T), \\
&= R_f + \gamma \mathrm{Cov}\left(\sum_{i=1}^{n} x_i R_i, R_T \right), \\
&= R_f + \gamma \mathrm{Cov}(R_T, R_T), \\
&= R_f + \gamma \mathrm{Var}(R_T).
\end{aligned}$$

Rearranging, we conclude

$$\gamma = \frac{\mathbb{E}(R_T) - R_f}{\mathrm{Var}(R_T)},$$

which implies

$$\mathbb{E}(R_i) = R_f + \frac{\mathbb{E}(R_T) - R_f}{\mathrm{Var}(R_T)} \mathrm{Cov}(R_i, R_T). \tag{13.1}$$

Suppose next that P is a portfolio made up of weights X_i for the assets with returns R_i. Then

$$\mathbb{E}(R_P) = \sum_{i=1}^{n} X_i \mathbb{E}(R_i),$$

and

$$\operatorname{Cov}(R_P, R_T) = \sum_{i=1}^{n} X_i \operatorname{Cov}(R_i, R_T).$$

Since the other terms in (13.1) do not change, we have

$$\mathbb{E}(R_P) = R_f + \frac{\mathbb{E}(R_T) - R_f}{\operatorname{Var}(R_T)} \operatorname{Cov}(R_P, R_T).$$

Therefore the expected return on any portfolio can be determined from its covariance with the tangent portfolio. We have completed step one of our argument.

We now turn to step two and suppose that the tangent portfolio is the same for every investor. Every investor will then hold a multiple of the risk-free asset and the same tangent portfolio. Since every asset must be owned by someone, the tangent portfolio is therefore the market portfolio; that is, every asset in the investment universe in the same proportions as they exist. Since the tangent portfolio is efficient, and it is the market portfolio, we have that the market portfolio is efficient.

Suppose then that every investor holds the same tangent portfolio, we then have for any portfolio P,

$$\mathbb{E}(R_P) = R_f + \frac{\mathbb{E}(R_M) - R_f}{\operatorname{Var}(R_M)} \operatorname{Cov}(R_M, R_P). \tag{13.2}$$

We set

$$\beta_P = \frac{\operatorname{Cov}(R_P, R_M)}{\operatorname{Var}(R_M)}. \tag{13.3}$$

This yields the CAPM equation:

$$\mathbb{E}(R_P) = R_f + \beta_P(\mathbb{E}(R_M) - R_f). \tag{13.4}$$

The CAPM says that the return on any portfolio is determined by its covariance with the market, and nothing else. In other words, we will receive no compensation for other sorts of risks. This completes step two of our argument.

For step three, we need to identify assumptions which suffice to guarantee that every investor has the same tangent portfolio. Typical assumptions are as follows.

- *Mean–variance investors.* Investors are assumed to only use mean–variance analysis to make their decisions. This was a crucial part of our derivation of the tangent portfolio. Alternatively, we could assume that asset returns are jointly normal so that means and variances describe everything.)
- *No tax effects.* Different investors are exposed to different tax rates. Often there are different rates for capital growth, and earned/saved income. These will affect different investors differently and so impact on investment decisions. We therefore assume tax is the same for everyone and for all types and amounts of income.
- *No transaction costs.* If there were transaction costs, then an investor's choice of investments would depend on existing holdings, since he would attempt to avoid incurring costs. These would vary from investor to investor and so we must assume transaction costs are nil.
- *Assets are divisible.* If every asset has to be held in the same proportion for each investor, then this is unlikely to result in integer holdings for all investors. We must therefore assume that they can be held in fractional amounts.
- *Not moving the market.* Whilst buying and selling shares moves market prices, we do not want any single investor's trading to move the market. This helps guarantee that all investors face the same decisions.
- *The risk-free asset.* Our derivation of the tangent portfolio and the argument that all investments would be a multiple of it assumed that there was infinite capacity to buy and sell at the risk-free rate. We must therefore assume that this holds.
- *Consistency of time horizon.* The mean returns and covariances of returns that determine the tangent portfolio depend upon the time horizon. We therefore must assume that the time horizon is the same for everybody.
- *Opinions.* The tangent portfolio depends on estimates of means and variances. For agreement on the tangent portfolio, these estimates must be the same.
- *All assets are marketable.* An investor's optimum portfolio will be affected by any assets he holds which cannot be marketed. For example, his income in his profession may have non-zero covariance with other investments. We therefore have to assume that even human capital is marketable and makes up part of the market portfolio.
- *Short sales allowed.* We recall that our derivation of the tangent portfolio does not require the weights to be positive. Hence we have to allow short sales. Interestingly, it turns out that the CAPM predicts that no-one short sells and accordingly, this axiom is not as necessary as it first appears: if

everyone is holding the same proportion of each asset then that proportion must be positive for every asset; someone has to own everything!

13.3 Assessing the CAPM assumptions

The assumptions of CAPM are many and highly dubious. We also know as a matter of practice that investors hold things other than the market portfolio. We assess a model based on the worth of its predictions not the merits of its assumptions and therefore the question we must ask is, "Does the CAPM equation hold in practice?" We can deconstruct this question into two parts:

- does the beta of a stock determine its expected return in a linear fashion; and
- do zero-beta portfolios have return equal to the risk-free rates?

We will return to these later in this chapter with some tests of CAPM.

13.4 Using CAPM

Importantly, we now must ask ourselves how we can make use of CAPM. From the CAPM equation (13.4) we see that it can be used to estimate the expected return of a security given the expected return of the market portfolio, the beta, and the risk-free rate. The key difference from the more general single-factor model is that there is no alpha. We illustrate this with the following example.

Example 13.1 The risk-free rate is 3. The stock's beta is 2. The expected return of the market is 10. To find the expected return on the stock we substitute into (13.4) to obtain

$$3+2(10-3)=17.$$ ◇

13.5 Implementing CAPM

Whilst the CAPM equation appears beautifully simple, we still need to determine the beta and, as with the single-factor model, we only ever have an estimate of beta. We also need to know the market portfolio and be able to estimate its expected return. This is no mean endeavour since the theoretical market portfolio contains every asset you could possibly invest in or even to which you could have exposure. It is not feasible for us to invest in every possible investment and a proxy is necessary. This is typically a market index such as the S & P 500 or the FTSE 100. Whilst convenient, this puts the CAPM model at even greater risk of being dubious.

Next suppose we form a portfolio of high beta stocks and ask ourselves what will happen. If CAPM is correct then we should get a return over a year which is the product of the portfolio with whatever the market returns over that year, plus a small error. Therefore, if the market goes up, we should make more, and if the market does down we should lose more. However, neither is guaranteed!

13.6 Eliminating the risk-free asset

Under the assumption that everyone is a mean–variance investor and other assumptions that guarantee that everyone has the same tangent portfolio which is necessarily the market, we derived the CAPM equation:

$$\mathbb{E}(R_P) = R_f + \beta_P(\mathbb{E}(R_M) - R_f),$$

with

$$\beta_P = \frac{\operatorname{Cov}(R_P, R_M)}{\operatorname{Var}(R_M)}, \tag{13.5}$$

for any portfolio P and M the market.

We used the risk-free asset in deriving CAPM. Suppose, however, that there is no risk-free asset in our market. Are we still able to derive something similar? Suppose that we take some efficient portfolio E with return R_E, then we know by our previous work that E is the tangent portfolio for a *hypothetical* risk-free rate S_f (see Chapter 4).

We note that our derivation of the expected return equation used nothing else about the risk-free rate. Hence repeating it, we get, for any portfolio P,

$$\mathbb{E}(R_P) = S_f + \frac{\mathbb{E}(R_E) - S_f}{\operatorname{Var}(R_E)} \operatorname{Cov}(R_P, R_E).$$

From the CAPM assumptions we have that:

- every investor is a mean–variance investor;
- every investor agrees on which portfolios are efficient.

We also know that:

- the efficient portfolios lie on a straight line in weight space;
- the minimal variance portfolio is efficient.

We therefore conclude that:

- every investor holds a portfolio lying on a straight line in weight space which starts at the minimal variance portfolio.

It follows that if we have a portfolio on this line and add to it the investments of another investor, then the resulting portfolio will also lie on this straight line. Therefore the portfolio of assets held by any group of investors taken together will be efficient. Suppose then that the group of investors is made up of all investors, then the resulting portfolio is simply the market portfolio. By our argument it must lie on the same line and hence we have shown that the market portfolio is efficient.

Since we have shown that the market portfolio is itself mean–variance efficient, then there exists some rate $S_{f,m}$ such that

$$\mathbb{E}(R_P) = S_{f,m} + (\mathbb{E}(R_M) - S_{f,m})\frac{\text{Cov}(R_P, R_M)}{\text{Var}(R_M)},$$

or

$$\mathbb{E}(R_P) = S_{f,m} + (\mathbb{E}(R_M) - S_{f,m})\beta_P.$$

Since S_f is just a hypothetical rate, we would like to eliminate it. To this end, we note that if a portfolio Z has zero beta, then

$$\mathbb{E}(R_Z) = S_{f,m}.$$

Hence, we have that $S_{f,m}$ is just the return on a zero-beta portfolio. Such portfolios will exist. To construct one we may use a combination of long and short positions to cancel out betas. In fact, there will be many such zero-beta portfolios. For concreteness, we will choose one, Z, that minimises variance. We then have, for any portfolio P, that

$$\mathbb{E}(R_P) = \mathbb{E}(R_Z) + \beta_P(\mathbb{E}(R_M) - \mathbb{E}(R_Z)).$$

This is called the two-factor version of CAPM, by reference to the fact that $\mathbb{E}(R_Z)$ provides another factor which may be varied.

As in CAPM, the expected return is a linear function of the β. The difference is that the straight line does not have to go through the risk-free rate. Importantly, we note that that the zero-beta portfolio will not generally have expected return equal to the risk-free rate.

We next show that Z is not efficient, by constructing a portfolio with lower variance and higher expected return than Z. To this end, consider E to be the minimal variance linear combination of Z and M. Since Z and M have zero covariance (Z has zero beta,) the holdings of Z and M are

$$\frac{\sigma_M^2}{\sigma_M^2 + \sigma_Z^2} \quad \text{and} \quad \frac{\sigma_Z^2}{\sigma_M^2 + \sigma_Z^2}.$$

Since it contains a positive amount of M, its beta is positive, and its return is greater than $\mathbb{E}(R_Z)$. Also, its variance is lower since it is the minimal variance combination.

13.7 Testing CAPM

In summary, CAPM makes the prediction that a stock's expected return is determined by its covariance with the market and lies on a straight line through the risk-free rate.

Two-factor CAPM makes the prediction that a stock's expected return is determined by its covariance with the market and lies on a straight line which does not have to be through the riskless rate but rather through the expectation of some zero-beta portfolio. We test our model by asking whether these predictions hold in practice.

The first problem with which we are confronted is that the CAPM is a statement about expectations. We cannot measure expectations. We can only measure returns. Returns are therefore typically used as a proxy for expectations. The second problem is estimation of the beta, an issue we encountered when studying single-factor models; we cannot measure future betas. We are reliant on measuring historical betas and using these as a proxy for the future beta. Implicit in this assumption is that betas are stable over time and furthermore, we are ignoring the noise in the beta estimation. Our third problem is that CAPM is a statement about the *market* portfolio. What does that actually mean? Market is supposed to mean that all possible assets that can be purchased. However, since it is not practical to work with that set, we can only test proxies. The proxy typically used is the all-share index. Taken together, this means that any test we devise will not be a test of CAPM directly but instead a test of CAPM jointly with the assumptions needed to make the test work. We list a few of these below by way of example.

- Historical returns are a good proxy for expectation.
- Historical betas are a good proxy for future beta.
- The S&P is a good proxy for the market portfolio.

Suppose then that we are a fund manager and we decide to use the CAPM. We would decide what level of risk multiplier to the market is appropriate and then form a portfolio of stocks with that level of beta. If CAPM were correct then the returns across time would be close to multiples of the market returns. We could look back historically to see if this prediction were borne out. Sharpe and Cooper [22] carried out such a test, which may be summarised as follows.

- Divide stocks into ten portfolios once a year according to their betas.
- Roll-over once a year for each level of betas.
- Use beta estimated from previous 5 years.
- Measure average return and average beta for each portfolio over a long period of time.

Their numbers lay close to a straight line with the equation

$$R_i = 5.54 + 12.75\beta_i;$$

this suggests a linear relationship between historical returns and betas. However, the risk-free rate across that period was 2. This suggests that whilst two-factor CAPM appears reasonable, one-factor CAPM does not look so good. Interestingly, they also found that in the long term, the investor who held the geometric-mean-maximising value of beta did better than the highest beta portfolio. This is consistent with the results of Chapter 10.

A standard methodology for testing CAPM and two-factor CAPM is the two-pass regression test. That is, one takes stock returns over several years. The returns are then regressed against market returns to estimate the betas, β_i, and the residual variances, $\mathrm{Var}(e_i)$. We then perform a second regression of the returns against a constant, the beta and the residual variances. We then have

$$R_i = a_1 + a_2\beta_i + a_3 \operatorname{Var}(e_i). \tag{13.6}$$

The essence of the CAPM is

$$a_1 = R_f, \tag{13.7}$$

$$a_2 = \mathbb{E}(R_M - R_f), \tag{13.8}$$

$$a_3 = 0, \tag{13.9}$$

so we can see if these actually hold. Note that such a test is "in sample" in the sense that the same data is used to estimate betas and to see how market returns related to them. Lintner carried out this test and found

$$a_1 = 0.108, \tag{13.10}$$

$$a_2 = 0.063, \tag{13.11}$$

$$a_3 = 0.237. \tag{13.12}$$

These are very different from CAPM: a_1 is way too big when compared to the risk-free rate; a_3 is very far from the zero predicted by both CAPM and two-factor CAPM.

Since then people have done further tests. Miller and Scholes, [13], suggest that Lintner's results actually arise from sampling bias coming from the residual risk, and that by using portfolios one can reduce this risk. They then find using portfolios that a_3 is close to zero as predicted. However, it is still the case that a_1 is not close to R_f. From these results CAPM looks bad, but the two-factor version looks somewhat better.

13.8 Roll's objection

Roll, [17], made a fundamental objection to the CAPM tests such as those we have outlined above. Using his logic, we may deduce that the tests impart no information as to the validity or otherwise of the CAPM model. His reasoning is as follows. Suppose we decide to test the two-factor CAPM and we use a portfolio, L, that turned out to be efficient as our proxy for the market portfolio. Then if we pick Z to have zero historical beta against L, the two-factor model derivation necessarily holds since we required no other assumptions in order to make it work. That is, since the two-factor CAPM's derivation holds, its predictions are correct by **construction.** Hence if we use a portfolio that turned out to be efficient as a proxy for the market portfolio, we have tested nothing at all. If we do not, then all we are testing is the efficiency of our proxy, not whether two-factor CAPM is a good portfolio model. It is important to realise Roll is not arguing anything about the validity of CAPM. Instead, he is arguing that the tests are meaningless.

One suggestion for overcoming the objection came from Shanken [21]. Since we only ever have a proxy and it is the efficiency of the proxy that is effectively being tested, he suggests studying the correlation between the proxy and the unknown market portfolio. In particular, he suggests testing the joint hypotheses that:

- the unobserved (true) market portfolio has greater than a given level of correlation (0.7) with the portfolio of stocks which we use as our proxy);
- that two-factor CAPM holds with respect to the unobserved market portfolio.

The essence of this approach is that if CAPM holds then there is a certain dependence between expected returns and beta with respect to the unknown market portfolio. It should follow that if our proxy is highly correlated with the market, then there must be a certain sort of dependence with that too. The upshot of Shanken's work is to reject this hypothesis statistically which suggests either that the market portfolio is very different from the proxy stock portfolio or that CAPM is wrong.

In assessing CAPM (or any other model) we must keep in mind that a statement is only meaningful if one can imagine circumstances under which it has been disproved. This is often referred to as the "Principle of Falsification" and forms the basis of scientific method which may be summarised as follows:

- a model predicts the results of new experiments;
- we do the experiments;
- if the experiments disagree with the model, then the model is wrong.

Whilst the principle is open to criticism, statements that fail it are certainly in a different category of usefulness than those that do not.

We conclude our chapter on CAPM by contrasting it with Tobin's separation theorem. This says that if **two** mean–variance investors have the same situation, except risk preferences, then they hold the same tangent portfolio but with differing amounts allocated between it and the risk-free asset. CAPM says that if **all** investors are mean–variance investors and **all** have the same situation then the tangent portfolio is the market portfolio. It follows that if we make people's situations vary even a little then CAPM does not hold. Tobin is only a statement about two investors.

13.9 Review

By the end of this chapter, the reader should be able to do the following questions and tasks.

1. What is the CAPM equation?
2. If we know the covariance of an asset with the tangent portfolio, how do we find its return? Give the derivation.
3. If everyone holds the same tangent portfolio, what can we say about its composition?
4. What are the assumptions of CAPM?
5. What do we typically use as the market portfolio when using CAPM?
6. Derive the two-factor CAPM equation.
7. Show that a zero-beta portfolio is not efficient in the two-factor CAPM.
8. What are the problems with testing CAPM?
9. What was Sharpe and Cooper's test of CAPM and what did they find?
10. What was Linter's test of CAPM and what did he find?
11. Explain Roll's objection to tests of the two-factor CAPM model.
12. What is the principle of falsification?

13. How did Shanken deal with the problems of testing CAPM and what did he find?

13.10 Problems

Question 13.1 Assuming that the risk-free rate is 9% and that the market portfolio has an expected return of 17%, what expected return would be consistent with the CAPM for a security with a beta of 1.5?

Question 13.2 Assume that the risk-free rate is 9%, and that the market portfolio has an expected return of 17% and a standard deviation of 20%. Under equilibrium conditions as described by the CAPM, what would be the expected return for a portfolio having no diversifiable risk and a standard deviation of 15%?

Question 13.3 If the risk-free rate is 3% and the market portfolio has an expected return of 9%, then, assuming the CAPM:

- what expected return is predicted for a security with a beta of 2?
- what expected return is predicted for a security with a beta of zero?

Question 13.4 The tangent portfolio for a mean–variance investor is known to have expected return 10 and standard deviation 10. The risk-free rate is 4. Another portfolio has standard deviation 20. What is its expected return if it has correlation 0.3 with the tangent portfolio?

Question 13.5 The tangent portfolio for a mean–variance investor is known to have expected return 10 and standard deviation 10. The risk-free rate is 5. Another portfolio X has standard deviation 20 and correlation 0.9 with the tangent portfolio. If an investor only wishes to hold X and the risk-free asset, how many units of X should be held to achieve a return of 20?

Question 13.6 Suppose that the two-factor CAPM model holds. Asset A has a beta of 0.5 and an expected return of 10. Asset B has a beta of 1.5 and an expected return of 20. What are the expected returns of the following portfolios and will they be mean–variance efficient?

- The market portfolio.
- A portfolio with zero beta.

Question 13.7 Suppose that the two-factor CAPM model holds. The market portfolio has expected return 15. A zero-beta portfolio has expected return 5. Give the expected returns of portfolios with:

- beta equal to 0.5;
- beta equal to 2.0.

Question 13.8 Suppose that the two-factor CAPM model holds. The market portfolio has expected return 15. A zero-beta portfolio has expected return 5. Give the betas of portfolios with:

- expected return 20;
- expected return 0.

Question 13.9 Suppose that both the two-factor CAPM model and the single-factor model hold. What can we say about the coefficients of the single factor model?

Question 13.10 Does the single-factor model imply CAPM? Does CAPM imply the single factor model? Can both hold at once? Can neither hold?

Question 13.11 Suppose that the two-factor version of CAPM holds. The risk-free rate is 3. The standard deviation of the market is 5. Stock A has standard deviation of 20 and correlation with the market of 0.25. Stock B has standard deviation of 10 and correlation with the market of 0.25. The expected return of A is 10 and of B is 7.5. What are the expected returns of:

- the market portfolio;
- a stock C with standard deviation of 8 and correlation with the market of 0.3?

Question 13.12 Suppose that the two-factor CAPM model holds. The market portfolio has expected return 10. A zero-beta portfolio has expected return 4. The risk-free rate is 1. Give the betas of portfolios not containing the risk-free asset with:

- expected return 20;
- expected return 0.

14

The arbitrage pricing model

14.1 Introduction

Arbitrage is a widely-used expression in the language of modern financial markets. It means opportunities for profit by finding and exploiting pricing anomalies between different securities and markets. In its proper sense, it refers to an opportunity to make a risk-free return over and above the risk-free rate. In a market where everything is functioning properly and all information is transmitted instantaneously to all participants, such opportunities should not exist. If they did, prices would move rapidly to eliminate the disparity. This gives us the principle of "no arbitrage," a very powerful tool for determining the prices of complex securities. It can also be made to apply to portfolio analysis. The idea is that if stock prices behave as in a multi-factor model but with no idiosyncratic risk, then by forming judicious combinations we can eliminate all risk. Once all the risk has been eliminated, a portfolio must return the risk-free rate. This in turn imposes conditions on portfolio returns.

14.2 Defining arbitrage

We start our analysis by formally defining arbitrage.

Definition 14.1 An *arbitrage* is a trading strategy in a portfolio of assets such that:

(1) the portfolio is initially of zero value;
(2) at some time $T \geq 0$ in the future, the portfolio has zero probability of having a negative value;
(3) and at the same time T, it has positive probability of positive value.

Simply stated, if we have an arbitrage, we can win but we can't lose.

There are many different sorts of arbitrage, the key examples of which are as follows.

(1) Instant arbitrage: we make an immediate profit at no risk.
(2) Static arbitrage: we buy and sell securities at time zero and then wait till some fixed T when we sell everything.
(3) Dynamic arbitrage: we continuously buy and sell assets according to what the market does.

The expression is also used in the context of "statistical arbitrage" where traders analyse statistical properties of securities and then trade in such a way as to make money on average. We note that this is not a true arbitrage in the scope of our definition, yet the term is widely used nonetheless. In this book, we will mainly be concerned with static arbitrage.

14.3 The one-step binomial tree

To motivate our analysis, we start with a simple example of a stock option.

- A stock is worth 100 today.
- It is worth 110 with probability 0.75 tomorrow.
- It is worth 90 with probability 0.25 tomorrow.
- Interest rates are zero and we can deposit and borrow with no limit.

How much is the right (but not the obligation) to buy the stock for 100 tomorrow worth? The option is called a *call* option *struck* at 100. The option is worth

$$110 - 100 = 10$$

with probability 0.75 and zero otherwise since it would not be used in the down-state.

The intuitively obvious answer is

$$0.75 \times (110 - 100) + 0.25 \times 0 = 7.5.$$

This turns out to be wrong! The correct answer is 5. By using the principle of no arbitrage, we can prove it.

14.4 The principle of no arbitrage

Simply stated, riskless profits are too good to be true. If such opportunities were to occur, then they would be immediately exploited through trading. The very action of trading to crystallise the profit would then move the market in such a way as to remove the arbitrage opportunity. We therefore assume

Definition 14.2 *The principle of no arbitrage*: in a market in equilibrium, there will be no arbitrage opportunities.

The principle of no arbitrage is very important and despite making only very weak assumptions has many powerful consequences. It is also called the "no free lunch" principle. In practical terms, this principle is used in conjunction with the prices of simple instruments to compute prices of complicated ones. This approach underlies most of derivatives pricing.

Closely related to the principle of no arbitrage, is a simple consequence: *the law of one price.* This says that if we can exactly synthesise the cash-flows of one portfolio with another portfolio then the two portfolios must have the same price today. To see that this follows from the principle of no arbitrage, suppose that it were not true. Then we would sell the more expensive portfolio, and buy the cheaper one. All cash-flows net against each other and we obtain a riskless profit, which would contradict the principle of no arbitrage and therefore cannot occur. This leads to the process of pricing by *replication.* That is, given a new security, we find a portfolio of simpler securities that replicates the security's cash flows. It follows that the price of the new security must be equal to the cost of creating the replicating portfolio.

14.5 Using replication to price a call option

If we can set up a portfolio that gives the same final value as another security no matter what happens, it is said to *replicate* the security. The value of the security and the replicating portfolio must then be the same by no arbitrage.

We can apply replication to our simple example of the stock option in Section 14.3. In particular, we have two possible states at the end: either the stock price goes up and the values of the stock, riskless deposit and option are respectively

$$110, \ 1, \ \text{and} \ 10;$$

or it goes down, and we get

$$90, \ 1, \ \text{and} \ 0.$$

We can write the asset values as a vector, with each place representing a final state. So the values are

$$\begin{pmatrix}110\\90\end{pmatrix},\ \begin{pmatrix}1\\1\end{pmatrix},\ \text{and}\ \begin{pmatrix}10\\0\end{pmatrix}.$$

We can write the option price as a linear combination of the other two assets:

$$\lambda\begin{pmatrix}110\\90\end{pmatrix}+\mu\begin{pmatrix}1\\1\end{pmatrix}=\begin{pmatrix}10\\0\end{pmatrix}.$$

We have two simultaneous equations in two unknowns. The solution is

$$\lambda = 0.5, \quad \mu = -45.$$

The portfolio of 0.5 stocks and −45 riskless deposits has final value 10 in the up-state and 0 in the down-state, as does the call option. The call option is therefore worth the same as the replicating portfolio. It must therefore be worth

$$0.5 \times 100 - 45 = 5.$$

Importantly, the final answer does **not** involve the probability of an up move. Suppose then that the probability of an up move is p. The expected value of the pay-off is $10p$. This gives us the right answer if and only if $10p = 5$; that is, if and only if $p = 0.5$. This leads us to an interesting property: if we compute expectations with $p = 0.5$, this expectation is the value of the option. We explore this further in our next section.

14.6 Risk-neutrality

We generally expect risky assets to have expected returns higher than riskless ones, or why we would be buy them? When the expected return of an asset for a given probability is the same as the riskless one, the probability is said to be *risk neutral.* Risk-neutral probabilities are intimately connected with the principle of no arbitrage. We illustrate by returning to our example where our investment universe consists of the stock, riskless deposit, and call option, and suppose that every asset's value today is equal to its expected value at time T. Suppose further that S is the stock price, B is the riskless deposit price, and C is the call option price. Then with $p = 0.5$, we have

$$S_0 = \mathbb{E}(S_1), \tag{14.1}$$

$$B_0 = 1 = \mathbb{E}(B_1), \tag{14.2}$$

$$C_0 = \mathbb{E}(C_1). \tag{14.3}$$

Any portfolio is a linear combination of these three assets. If we hold α units of S, β units of B and γ units of C, and our portfolio is of initial value zero, then

$$\begin{aligned}\mathbb{E}(\alpha S_1 + \beta B_1 + \gamma C_1) &= \alpha S_0 + \beta B_0 + \gamma C_0, \\ &= 0.\end{aligned} \tag{14.4}$$

The risk-neutral expectation of an arbitrage portfolio is positive: positive probability of being positive and zero probability of being negative. Hence, no arbitrages can occur if, for all portfolios A, we have

$$\mathbb{E}(A_1) = A_0.$$

However, we recall that our real-world probability p actually equalled 0.75. We saw that an arbitrage for one p was however, an arbitrage for all p with $0 < p < 1$, since the definition of arbitrage only used the sets of zero probability. Hence, if we have no arbitrage with $p = 0.5$, we have no arbitrage with $p = 0.75$ either. This gives us an approach to pricing.

First, we change probabilities so that for every asset A we know the price of, we have

$$\mathbb{E}(A_1) = A_0.$$

Then for any security we wish to price, we *define*

$$C_0 = \mathbb{E}(C_1).$$

We are then guaranteed no arbitrage.

14.7 Interest rates and discounting

In the above analysis we assumed that there were no interest rates. This is somewhat unrealistic. If there are interest rates, the riskless asset will grow in value across time and we then have $B_1 \neq B_0$. Since B is riskless, we have, whatever we do to our probabilities,

$$\mathbb{E}(B_1) = B_1 \neq B_0.$$

The solution to this problem is to discount everything. That is, we first divide at all points by the riskless asset B, and find that

$$\mathbb{E}(B_1/B_1) = \mathbb{E}(1) = 1 = B_0/B_0.$$

We then find probabilities such that

$$\mathbb{E}(S_1/B_1) = S_0/B_0.$$

The stock then still has the same expected return as the riskless deposit. We examine this by returning to our call option example.

Example 14.3 Suppose that the risk-free rate is now 5%. We have two possible states at the end of the time period. Either the stock price goes up and the values of the stock, riskless deposit and option are respectively

$$110,\ 1.05,\ \text{and}\ 10;$$

or the stock goes down and we get

$$90,\ 1.05,\ \text{and}\ 0.$$

Since the interest rate is 5%, the riskless deposit grows in value. For replication, we now have

$$\lambda \begin{pmatrix} 110 \\ 90 \end{pmatrix} + \mu \begin{pmatrix} 1.05 \\ 1.05 \end{pmatrix} = \begin{pmatrix} 10 \\ 0 \end{pmatrix}.$$

As before, we get $\lambda = 0.5$, but now

$$\mu = -45/1.05 = -42.86.$$

The call option value is therefore

$$100 \times 0.5 - 42.86 = 7.14.$$

Introducing interest rates has increased the value of the call option.

Alternatively, we may try the risk-neutral approach. That is, we need to find the probability p such that

$$\frac{90}{1.05}(1-p) + \frac{110}{1.05}p = 100,$$

or

$$90(1-p) + 110p = 105.$$

We solve this to get $p = 0.75$, which, by coincidence, is where we started. We compute

$$\begin{aligned} C_0 &= B_0 \mathbb{E}(C_1/B_1), \\ &= 1.(0.75 \times 10/1.05 + 0.25 \times 0), \\ &= 7.14, \end{aligned} \tag{14.5}$$

which is the same as we obtained using the replication approach. ◇

14.8 The trinomial tree and limitations of no arbitrage

The no arbitrage principle, whilst powerful, is not without its limitations. We illustrate this with the example of a stock with initial value 100, which can take three values, $90, 100$ and 110, at the end of some period in a no-interest-rate environment. As before, our objective is to price a call option struck at 100. We now have three states to take into account. The value of the assets (stock, deposit and option, respectively) in the these three states are as follows:

$$\begin{pmatrix} 110 \\ 100 \\ 90 \end{pmatrix}, \begin{pmatrix} 1 \\ 1 \\ 1 \end{pmatrix} \text{ and } \begin{pmatrix} 10 \\ 0 \\ 0 \end{pmatrix}.$$

By attempting to follow our process in Section 14.5 above, we find ourselves with three equations in two unknowns. This is not solvable and hence we can no longer replicate. If we solve in the top and bottom states, we get the same as before: $\lambda = 0.5$, $\mu = -45$. However, when $S = 100$, this gives 5 rather than 0. This says that we can construct a portfolio that *super-replicates* for 5; that is, we have a portfolio that always pays the same or more than the call option. Using no arbitrage, this says that the call option is worth less than 5. Otherwise, we trade the super-replicating portfolio minus the call option and have an arbitrage. However, we have that the call option is always non-negative and sometimes positive. Therefore, we can *sub-replicate* with zero and hence the call option is of positive value. We have bounded the call option between zero and five. We would like to show this is optimal.

Attempting the risk-neutral approach, we need to find the probabilities for which

$$\mathbb{E}(S_1) = S_0.$$

One can see by symmetry that this requires

$$p_{\text{up}} = p_{\text{down}}.$$

Since probabilities must add up to less than 1 and be positive, we have

$$0 < p_{\text{up}} = p_{\text{down}} < 0.5.$$

The value of our call option is therefore

$$10 p_{\text{up}},$$

and this will be between 0 and 5. The risk-neutral argument says that these prices are not arbitrageable. The replication argument said that all other prices were. So the set of non-arbitrageable prices is the numbers between 0 and 5.

The moral of this example is that whilst the principle of no arbitrage sometimes gives a unique price, often it will only yield bounds on the price.

14.9 Arbitrage and randomness

We next apply the principle of no arbitrage to demonstrate that the risk-free rate is unique. That is, suppose there are two portfolios X and Y which return the non-random (and hence risk-free) rates r_X and r_Y respectively. We show that these are the same by reasoning as follows.

If $r_X > r_Y$, then we form the portfolio of:

- \$1 of X
- minus \$1 of Y;

this has value after a year of

$$1 + r_X - 1 - r_Y = r_X - r_Y > 0,$$

giving an arbitrage. Hence $r_X \leq r_Y$. By symmetry, we conclude that $r_X = r_Y$.

For any interesting asset, the returns are random. However, it is sometimes possible to eliminate randomness by holding a combination of long and short positions in such a way to as to cancel it out. We illustrate this in the next example.

Example 14.4 Suppose that X is a random variable and that the risk-free rate is r_f. Suppose further that asset A returns

$$r_A + X,$$

and asset B returns

$$r_B + 2X,$$

for $r_A, r_b \in \mathbb{R}$, with the *same* random variable X.

Next suppose we create a portfolio, P, consisting of 2 units of A and -1 units of B. The return on this is given by

$$2r_A + 2X - r_B - 2X = 2r_A - r_B.$$

We note that this is an *actual* return, **not** an *expected* return. Using the uniqueness of risk-free assets we have

$$2r_A - r_B = r_f,$$

and, equivalently,

$$2(r_A - r_f) = r_B - r_f. \qquad \diamond$$

One interpretation of this result is that asset B has twice the exposure to the risk factor X and so needs twice the excess return to compensate for it. This is the fundamental idea which underpins the arbitrage pricing theory or "APT"

14.10 Arbitrage Pricing Theory

We develop our theory by returning to the multi-factor model which we introduced in Chapter 6. We take a number of uncorrelated indices I_j for $j = 1,\ldots,L$ which are random and set

$$R_i = a_i + \sum_{j=1}^{L} b_{ij} I_j + c_i.$$

for $i = 1,\ldots,N$. The numbers a_i and b_{ij} are constants. The c_i are random variables with zero mean and are uncorrelated with the indices.

Suppose next that there were no idiosyncratic terms in the multi-factor model; that is,

$$c_i = 0 \quad \text{for all } i,$$

and that N is much bigger than L. We can apply the principle of no arbitrage to the returns. There are L risk factors, but we have N assets with N much bigger than L. When the number of risk factors is less than the number of assets, it is a general rule that no arbitrage will reduce the set of possible prices. In particular, it means we can create riskless portfolios from a collection of risky assets. Since all riskless portfolios should return the same amount, we must have relations between the possible values of the loadings on the risk factors.

Taking the case where $L = 2$, this will imply that there exist terms, μ_i, such that

$$\mathbb{E}(R_i) = \mu_0 + \mu_1 b_{i1} + \mu_2 b_{i2}. \tag{14.6}$$

The terms b_{il} are like generalised betas. They reflect the extra compensation achieved for taking risk on the factor I_l.

Next suppose we have two portfolios, P, Q, which have the same loadings (for the same set of risk factors I_j); that is,

$$b_{pj} = b_{qj}.$$

We want to prove that they have the same expected returns; that is, we want to prove the following theorem.

Theorem 14.5 *In the arbitrage pricing model (assuming no arbitrages exist),* *if two portfolios have no idiosyncratic risk and the same factor loadings, they have the same expected return.*

Proof We show that if they have different expected returns then an arbitrage exists. The argument is similar to that which we used to show the uniqueness of risk-free rates.

Suppose that

$$\mathbb{E}(R_P) > \mathbb{E}(R_Q),$$

since otherwise we may swap P and Q. We construct a portfolio comprising a long position in P and a short position in Q. Since they have the same randomness, the difference is deterministic (i.e. riskless) and will be an arbitrage.

Accordingly, consider the portfolio, A, consisting of 1 unit of P worth \$1 and -1 units of Q worth $-\$1$. This portfolio has zero initial value. The expected value of P after a year is

$$\mathbb{E}(1+R_P) = 1+\mathbb{E}(R_P).$$

The expected value of Q after a year is

$$\mathbb{E}(1+R_Q) = 1+\mathbb{E}(R_Q).$$

Therefore the expected value of A after a year is

$$\mathbb{E}(R_P - R_Q) > 0.$$

Furthermore, the actual value after a year is $R_P - R_Q$ which equals

$$a_P + \sum b_{Pi} I_i - a_Q - \sum b_{Qi} I_i = a_P - a_Q.$$

The value of $P - Q$ after a year is deterministic and has positive expectation, and so is a fixed positive number. We conclude that it will always be positive after a year and that the portfolio $A = P - Q$ is an arbitrage. Since a difference in expected returns implies an arbitrage, the principle of no arbitrage tells us that expected returns must be the same for P and Q. □

It can be shown, using these results, that there exist constants μ_j such that if

$$R_i = a_i + \sum_{j=1}^{L} b_{ij} I_j,$$

then

$$\mathbb{E}(R_i) = \mu_0 + \sum_{j=1}^{L} b_{ij} \mu_j.$$

14.11 Computations

A typical problem is to determine the constants μ_j and then apply them to compute the expected return for another portfolio. Accordingly, suppose we have $L+1$ constants to determine. To do this, we will require the expected returns and b_{Pj} terms for $L+1$ portfolios.

We note that if we are given a riskless rate r_f then we necessarily have

$$\mu_0 = r_f,$$

and we require a further L portfolios to determine $\mu_1, \dots, \mu_L$. The terms b_{Pj} are sometimes called *loadings*.

We draw attention to the distinction between weights and loadings. Weights must sum to 1 and express how much of each asset is held, whereas loadings can sum to anything and express the exposure to each risk factor.

If we have $L+1$ assets and L factors in our multi-factor model, then we can generally (but not always) construct any set of weights, b_{pj}, on the factors. We note that in order to pin down the risk-free return we need $L+1$ assets rather than just L. The expected return for that set of weights is then simply the linear combination of the expected returns of the individual assets. We illustrate this in a situation for which $L = 2$ and $N > 3$, and show that we can deduce the characteristics of the fourth and later assets from the first three.

Example 14.6 We have three assets as follows.

Asset	b_{i1}	b_{i2}	Expected return,
A	1	2	10,
B	3	1	12,
C	2	2	11,
D	x	y	z.

We first show how to find the return on any portfolio with exposure to either the risk-free asset or just one risk factor. We start by solving for the risk-free return. That is, we want to find α, β, γ holdings of A, B and C with total weight 1, and such that all exposure to the risk factors I_1 and I_2 is eliminated. This yields the following system of linear equations:

$$\alpha + \beta + \gamma = 1; \tag{14.7}$$

$$\alpha + 3\beta + 2\gamma = 0, \quad \text{for } b_{P1}; \tag{14.8}$$

$$2\alpha + \beta + 2\gamma = 0, \quad \text{for } b_{P2}. \tag{14.9}$$

We can rewrite these conditions as a matrix equation:

$$\begin{pmatrix} 1 & 1 & 1 \\ 1 & 3 & 2 \\ 2 & 1 & 2 \end{pmatrix} \begin{pmatrix} \alpha \\ \beta \\ \gamma \end{pmatrix} = \begin{pmatrix} 1 \\ 0 \\ 0 \end{pmatrix}.$$

The solution to this equation is

$$\begin{pmatrix} 4 \\ 2 \\ -5 \end{pmatrix}.$$

The return on the riskless asset is found by multiplying these weightings by the asset returns for A, B, and C and summing; that is,

$$4 \times 10 + 2 \times 12 - 5 \times 11 = 9.$$

We can check our solution by substitution to confirm that the portfolio's loading on the first factor is

$$4 \times 1 + 2 \times 3 - 5 \times 2 = 0$$

and on the second factor is

$$4 \times 2 + 2 \times 1 - 2 \times 5 = 0$$

as expected.

Next suppose that a portfolio has a loading of 1 on the first factor and zero on the second. We want to know its expected return. To this end we have

$$\begin{aligned} \alpha + \beta + \gamma &= 1, \\ \alpha + 3\beta + 2\gamma &= 1, \quad \text{for } b_{P1}, \\ 2\alpha + \beta + 2\gamma &= 0, \quad \text{for } b_{P2}. \end{aligned}$$

As a matrix equation, we get

$$\begin{pmatrix} 1 & 1 & 1 \\ 1 & 3 & 2 \\ 2 & 1 & 2 \end{pmatrix} \begin{pmatrix} \alpha \\ \beta \\ \gamma \end{pmatrix} = \begin{pmatrix} 1 \\ 1 \\ 0 \end{pmatrix}.$$

The solution is

$$\begin{pmatrix} 3 \\ 2 \\ -4 \end{pmatrix}.$$

The portfolio's expected return is

$$3 \times 10 + 2 \times 12 - 4 \times 11 = 10,$$

and we can substitute to confirm that the loading on the first factor is

$$3 \times 1 + 2 \times 3 - 4 \times 2 = 1$$

and the loading on the second factor is

$$3 \times 2 + 2 \times 1 - 4 \times 2 = 0,$$

as required.

Next suppose that a portfolio P in assets A, B and C has loading of 0 on I_1 and 1 on I_2 and that we want to find its expected return. Accordingly, we want to find α, β, γ as follows:

$$\begin{aligned} \alpha + \beta + \gamma &= 1, & \\ \alpha + 3\beta + 2\gamma &= 0, & \text{for } b_{P1}, \\ 2\alpha + \beta + 2\gamma &= 1, & \text{for } b_{P2}. \end{aligned}$$

As before, this may be expressed as a matrix equation:

$$\begin{pmatrix} 1 & 1 & 1 \\ 1 & 3 & 2 \\ 2 & 1 & 2 \end{pmatrix} \begin{pmatrix} \alpha \\ \beta \\ \gamma \end{pmatrix} = \begin{pmatrix} 1 \\ 0 \\ 1 \end{pmatrix}.$$

The solution is

$$\begin{pmatrix} 3 \\ 1 \\ -3 \end{pmatrix}.$$

We then have that the return on the second factor with unit weight is

$$3 \times 10 + 1 \times 12 - 3 \times 11 = 9.$$

As required, the loading on the first factor is

$$3 \times 1 + 1 \times 3 - 3 \times 2 = 0,$$

and the loading on the second factor is

$$3 \times 2 + 1 \times 1 - 3 \times 2 = 1.$$

◇

Next suppose we have weights b_{p1} and b_{p2} on the two factors. What is the expected return in this more general case? To solve this, we may use the fact that we have already solved for the risk-free rate and constructed portfolios with sole exposure to each of the first factor I_1 and the second factor I_j. We can then construct the replicating portfolio as

$$\begin{aligned} &b_{p1} \text{ units of the first factor portfolio,} \\ &b_{p2} \text{ units of the second factor portfolio,} \\ &1 - b_{p2} - b_{p1} \text{ units of the riskless portfolio.} \end{aligned}$$

The expected return on the general portfolio will be

$$b_{p1}\mathbb{E}(R_{\text{first}}) + b_{p2}\mathbb{E}(R_{\text{second}}) + (1 - b_{p1} - b_{p2})\mathbb{E}(R_{\text{riskless}}). \tag{14.10}$$

To compute this, we can either first construct the returns for each of the portfolios with sole exposure to one risk factor, or we can proceed directly.

Example 14.7 With the above assumptions, what is the expected return on a portfolio with exposure of 2 to I_1 and 1 to I_2? To compute this we substitute into (14.10) to obtain expected return

$$2 \times 10 + 1 \times 9 + (1 - (2+1)) \times 9 = 20 + 9 - 18 = 11. \qquad \diamond$$

Returning to equation (14.6), we may substitute our return results for each of the risk-free rate, the first-factor only and the second-factor only portfolios to obtain

$$\mu_0 = 9, \quad \mu_1 = 1, \quad \mu_2 = 0.$$

To further illustrate our methodology, we consider an example where there is a risk-free asset.

Example 14.8 Suppose a risk-free asset, F, returns 5 annually, and we have two assets A and B as follows.

Asset	b_{i1}	b_{i2}	Expected return,
A	1	0.5	11,
B	0.5	1	12,
C	x	y	z.

Let the gain for holding one exposure to b_{p1} be μ_1, and to b_{p2} be μ_2; then the expected returns on A and B are

$$\begin{aligned} 5 + \mu_1 + 0.5\mu_2 &= 11, \\ 5 + 0.5\mu_1 + 1\mu_2 &= 12. \end{aligned}$$

We therefore have two simultaneous equations in two unknowns:

$$\begin{aligned} 1\mu_1 + 0.5\mu_2 &= 6, \\ 0.5\mu_1 + 1\mu_2 &= 7. \end{aligned}$$

This has the solution

$$\mu_1 = \frac{10}{3}, \qquad \mu_2 = \frac{16}{3}.$$

An asset with x units of exposure to I_1 and y units of exposure to I_2 will therefore have expected return given by

$$5 + \frac{10}{3}x + \frac{16}{3}y. \qquad \diamond$$

14.12 An alternative approach to computation

Whilst the above approach is intuitive, there is another quicker method when it comes to practical computation. We illustrate this here for the same assets. That is, we have

Asset	b_{i1}	b_{i2}	Expected return,
A	1	2	10,
B	3	1	12,
C	2	2	11.

This implies that the following equations hold:

$$\begin{aligned} \mu_0 + 1\mu_1 + 2\mu_2 &= 10, \\ \mu_0 + 3\mu_1 + 1\mu_2 &= 12, \\ \mu_0 + 2\mu_1 + 2\mu_2 &= 11. \end{aligned}$$

We have three equations in three unknowns. Hence we are able to solve them for μ_0, μ_1, μ_2 using a matrix equation. That is, the system is equivalent to

$$\begin{pmatrix} 1 & 1 & 2 \\ 1 & 3 & 1 \\ 1 & 2 & 2 \end{pmatrix} \begin{pmatrix} \mu_0 \\ \mu_1 \\ \mu_2 \end{pmatrix} = \begin{pmatrix} 10 \\ 12 \\ 11 \end{pmatrix}.$$

This has the unique solution

$$\begin{pmatrix} 9 \\ 1 \\ 0 \end{pmatrix}.$$

This second approach is notably easier. However, the first approach demonstrates to us why the method works.

14.13 Introducing realism

Returning now to the assumptions underlying our APT model, we must recognise that even if we believe the market to be driven by $L < N$ common risk factors, each asset will have an idiosyncratic part to its return. That is, in the expression

$$R_i = a_i + \sum_{j=1}^{L} b_{ij} I_j + c_i,$$

we would not expect to have $c_i = 0$ for individual securities. We must therefore conclude that the no arbitrage argument does **not** truly apply to this situation. That said, whilst individual securities do have residual risk, it is reasonable to assume that this is effectively zero for a well-diversified large portfolio. It follows that we can reasonably apply the results to large portfolios but not to individual stocks. However, whilst no arbitrage is not powerful enough to make the APT apply for individual stocks, it is often used nevertheless. One possible argument to support this approach is that the idiosyncratic terms are diversifiable and hence should not attract risk premia. A detailed discussion of the issues regarding what is required to deduce what, including a convincing argument that no arbitrage is not enough can be found in Cochrane, [3].

14.14 APT versus CAPM

APT and CAPM are separate theories but they do not contradict each other. CAPM instead places constraints on the values of parameters in APT. For example, if we have a two-factor APT, then the expected return on asset i satisfies

$$\mathbb{E}(R_i) = R_f + b_{i1}\mu_1 + b_{i2}\mu_2,$$

for some μ_1, μ_2.

The CAPM places constraints on μ_1, μ_2. That is, we know we can construct portfolios with one unit of exposure to a particular index I_j. Let P_j be such a portfolio. We then have from the APT that

$$\mathbb{E}(P_j) = R_f + \mu_j.$$

We also have from CAPM that

$$\mathbb{E}(P_j) = R_f + \beta_j(\mathbb{E}(R_M) - R_f),$$

with β_j the covariance between the market and P_j. That is, CAPM simply requires that

$$\mu_j = \beta_j(\mathbb{E}(R_M) - R_f),$$

so the two theories are not contradictory. Importantly, however, APT gives us more room for manoeuvre. It can certainly be correct without the CAPM being correct. It is also possible for CAPM to hold without APT since CAPM makes no assumptions on the number of factors driving movements.

14.15 APT in practice

Suppose that a fund manager has information relating only to the market and its historical movements. The problem is to identify the underlying risk factors for APT. There are two basic approaches to identifying these indices:

1. Use economic indicators, e.g. inflation, interest rates, risk premia, consumer confidence, production. (These would have to be made uncorrelated.)
2. Attempt to statistically extract unobservable factors.

This also relates to testing APT. That is, APT says that if returns are given by an L-common-factor model [plus an idiosyncratic part], then the expected returns satisfy certain relations. It does not say what the common factor model is. Our test therefore has to identify the putative factors, as well as examine dependence on them.

If we believe that we know the set of true indices, testing becomes easier. Chen, Roll, Ross [2] examined APT using the indices of:

- inflation;
- long-term interest rates minus short-term interest rates;
- spread between risky and riskless bonds;
- industrial production.

We examine each of these factors in turn below.

- **Inflation**: inflation affects the level of interest rates and their difference remains fairly constant. Inflation also affects the size of future cash flows in nominal terms, (but not so much in *real* terms). Furthermore, there is a general belief that high inflation spells economic bad times.
- **Long-term minus short-term interest rates**: the difference in these rates affects the value of short-term cash flows after discounting versus long-term cash flows after discounting. Additionally, the shape of the yield curve reflects economic expectations with a downwards-sloping curve generally believed to reflect upcoming recession, since it indicates that short-term rates will be high in order to rein in inflation in times of growth, but in the long term it will be low to encourage growth.

- **The bond spread**: this is affected by the estimated probability of defaults and the risk aversion of investors. Both of these will directly impact stock prices.
- **Industrial production**: a fairly clear indicator of economic outlook.

Chen, Roll and Ross found that each of these four factors had strong effects on stock market returns with industrial production having the biggest. They also found, perhaps surprisingly, that for the period they studied, oil price was not a significant factor.

An alternative approach is to use principal components analysis to attempt to statistically determine how many factors affect expected returns. Tests seem to suggest between three and five factors are significant. This is more than one would expect from CAPM!

14.16 Applications of APT

One use of APT is in index tracking. Some funds have the objective of returning to the investor the growth in a particular market index. The fund therefore needs to synthesise this return which, for a large index, involves many different stock positions. One would prefer to use fewer stocks. The APT approach can tell us how to achieve market returns with a smaller number of stocks.

Another use is to achieve exposure to a quantity on which we have a view. For example, if we were of the view that inflation is set to increase, then we could use APT to construct a portfolio which is sensitive to inflation but not the other main factors. If our view is correct, we will make money.

14.17 Criticising APT

Factor analysis will always produce something, no matter what the inputs. For computational reasons, tests have generally been done with groups of stocks rather than individual ones. However, to obtain the true diversification effects touted by the APT one should be using all stocks individually. When testing, we also have to do the following as discussed for CAPM:

- use actual returns as a proxy for expected returns;
- estimate exposures to factors using historical data.

14.18 Review

By the end of the chapter the reader should be able to

1. state the principle of no arbitrage;
2. state the "law of one price";
3. state the assumptions for the APT and prove that under these the factor loadings determine returns;
4. explain to what extent the CAPM and the APT are compatible;
5. explain which indices Chen, Roll and Ross used for their work and why, summarise their findings, and discuss whether their results are compatible with CAPM.

14.19 Problems

Question 14.1 The risk-free rate is 3 and the asset X returns $5+Z$ for some random variable Z. Compute α_j if asset Y_j returns

$$\alpha_1 + 2Z, \quad \alpha_2 - Z, \quad \alpha_3 + 1.5Z.$$

Question 14.2 If asset A returns $3+Z$ and asset B returns $7+2Z$ for some random variable Z, then what is the risk-free rate?

Question 14.3 For each of the following state whether there's an arbitrage, what sort of arbitrage is being claimed, and discuss whether it really is an arbitrage.

- A share in company X carries the right to buy a container of cocoa beans for \$100. A container trades for \$200 on the market, and the current market price of shares in X is \$60.
- Three mathematicians claim that by continuously trading in a stock and riskless bonds they can replicate the pay-off of a call option for less than its market price.
- An econometrician observes that if a particular stock outperforms the market three days running then on the fourth day it underperforms 90% of the time, and suggests short selling for such days.

Question 14.4 Suppose that we are given

Asset	b_{i1}	b_{i2}	Expected return,
A	2	1	13,
B	3	1	14,
C	1	2	20.

Find the risk-free rate, and the expected return per unit of risk for each of I_1 and I_2.

Question 14.5 Suppose that we are given

Asset	b_{i1}	b_{i2}	Expected return,
A	2	1	20,
B	1	2	14,
C	0	1	3.

Find the risk-free rate, and the expected return per unit of risk for each of I_1 and I_2.

Question 14.6 Suppose that we are given

Asset	b_{i1}	b_{i2}	Expected return,
A	3	1	28,
B	1	2	22,
C	2	–1	16.

Find the risk-free rate, and the expected return per unit of risk for each of I_1 and I_2.

Question 14.7 Suppose that the following two-index model describes asset returns:

$$R_i = a_i + b_{i1}I_1 + b_{i2}I_2 + e_i,$$

with a_i, b_{i1}, b_{i2} constants and the random variables I_1, I_2, e_i uncorrelated. We also assume $\mathbb{E}(e_i) = 0$. Suppose the riskless rate is 5. Suppose further that there are well-diversified portfolios with the following characteristics:

Portfolio	Expected Return	b_{i1}	b_{i2}
A	10	0.5	0.5
B	12	1.2	0.2

What is the expected return on a well-diversified portfolio with $b_{i1} = 0.2$, and $b_{i2} = 0.8$? What is the expected return on a well-diversified portfolio with $b_{i1} = 0.8$, and $b_{i2} = 0.2$?

Question 14.8 Suppose that the following two-index model describes asset returns:

$$R_i = a_i + b_{i1}I_1 + b_{i2}I_2 + e_i,$$

with a_i, b_{i1}, b_{i2} constants and the random variables I_1, I_2, e_i uncorrelated. We also assume $\mathbb{E}(e_i) = 0$. Suppose the riskless rate is 4. Suppose further that there are well-diversified portfolios with the following characteristics:

Portfolio	Expected Return	b_{i1}	b_{i2}
A	10	0.75	0.25
B	18	2	2

What is the expected return on a well-diversified portfolio with $b_{i1} = 1$, and $b_{i2} = -1$? What is the expected return on a well-diversified portfolio with $b_{i1} = 0.5$, and $b_{i2} = 0.5$?

15
Market efficiency and rationality

15.1 Introduction

In our analysis so far we have proceeded on the basis that investment decisions are made by measuring characteristics of asset prices in terms of the distribution of their returns. We have not paused to examine the fundamental pricing of stocks and to ask ourselves whether stocks are indeed correctly priced. The reason for this is that we are tacitly assuming the concept of *market efficiency* – that it is not possible to make money by exploiting mispricings in the market – together with the related concept of *market rationality* – that stock prices are accurately priced by the market after taking into account risk adjustments. In this chapter, we examine these concepts and discuss to what extent they are correct. The reader should remain aware however, that this is an area about which debate rages with something of a mystical fervour. For every argument countering efficiency, an opposing argument will emerge to explain why this does not constitute a true counterexample and consequently modern finance literature abounds with lively debate between the believers and sceptics in market efficiency.

15.2 Defining efficiency

The concept of stock market efficiency arose from interesting observations made in the 1960s that actively-managed funds did not appear to outperform passively-managed funds. In an actively-managed fund the fund manager selects specific investments that he believes will do well, a practice sometimes referred to as "stock-picking." By contrast, a passively-managed fund will purchase a portfolio that well represents the general market, or a sector thereof, and will do little more. Examples of these are index-tracker funds. Active funds generally charge more than passively-managed ones for the extra work and the perceived star quality of the fund manager. However, if a high-fee active fund fares no better than a low-fee passively-managed fund, why bother? At this point, it is worth remembering that most stocks are held by funds. Accordingly, the fact that the average fund manager does not outperform the market is therefore arithmetically inevitable. However, if our story is that some fund managers are good and some are poor, then one would expect a high degree of correlation from year to year as to which actively-managed funds do well; this does not appear to be the case.

Market efficiency seeks to encapsulate the fundamental idea that there is no advantage to being a "clever" fund manager. That is, there is no trading strategy that will result in statistically significant long-term outperformance. To caputre this idea, the various definitions of efficiency are usually phrased in terms of information. Information about a firm might include that which is available to the public, e.g. published accounts and stock exchange announcements, or private information such as that which might be held in confidence by senior executives of the firm. Information might also relate to external factors such as the global economy which in turn may influence either the fundamentals of the business or investor demand for asset. To consistently outperform, a fund manager must deploy strategies superior to his rivals. In short, he must know or understand something that others do not.

Statements of market efficiency are generally classified into three types:

- Strong: all information, public or private, is already reflected in stock prices.
- Semi-strong: all public information is already reflected in stock prices.
- Weak: all information in historical prices is already reflected in stock prices.

The use of the word *reflected* is important here and means that we cannot use the information to make excess returns. One might be able to make extra money by taking on risk: market efficiency does not say that one cannot gain extra returns by taking on risk nor does it rule out luck. It is also important to note that whilst efficiency requires the relevant information to be reflected in the asset price, it does not preclude other factors affecting the price. Asset prices may still fluctuate with an element of random noise in an efficient market. However, such truly random variation cannot be used to drive consistent outperformance.

We next analyse the three concepts of efficiency in turn and explore to what extent they model observed the observed market.

First, suppose that markets are strongly efficient. This says that no information we come by, no matter how confidential, will help us make extra money.

For example, your friend who works for MegaBank tells you that they will be making a takeover of MiniBank tomorrow and this will drive the price of MiniBank up. In a strongly efficient market, this will not help you make money. Lawmakers certainly do not believe in strong efficiency of stock markets: insider trading, such as in the above hypothetical example, is a crime in most countries. If markets were indeed strongly efficient, there would be no need for such prohibition since there would be no benefit to derive from it!

Next suppose we live in a world where markets are semi-strong. Insider trading would still be profitable, but publicly available information would have no potential use. Consequently, stock-picking would be a waste of time and there is no advantage in being intelligent beyond diversification to reduce risk. A corollary, however, is that there is no disadvantage in being stupid! Returns should be truly random.

Finally, we turn to the third concept of weak efficiency. This asserts that you cannot make money by doing mathematical or statistical analysis of stock price movements and hence hedge funds are wasting their time if they employ mathematicians to analyse stock price movements. For example, yesterday's price movement should not affect today's. This is referred to as zero autocorrelation and, given a time series of changes

$$\Delta S_i = S_{i+1} - S_i,$$

the two series

$$\Delta S_{i+1} \quad \text{and} \quad \Delta S_i$$

should be uncorrelated.

We pause momentarily in our analysis to make some basic philosophical observations about market efficiency. Firstly, we should not lose sight of the fact that an efficient market, however defined, is a model of real-world markets and no more. As with all models, the question is how well empirical evidence supports the assumptions of the model. When it comes to market efficiency, rather than asking whether a market is efficient, a better question might be how efficient? Interestingly, we observe, as a matter of logic, that a market cannot be perfectly efficient or at least that enough market participants must not believe in perfect efficiency for an efficient state to be approximated. To see this, suppose that markets were perfectly efficient and that all information (however defined) is already reflected in market prices. Then it would be worth no-one's time or expense to seek out and process information since no value could be had from such endeavours. However, we arrive at a contradiction since with no-one processing information, there is now no mechanism for new information to be incorporated into asset prices. It follows then that market participants have to believe in inefficiencies for an efficient state to be approached. One might then argue that inefficiencies, where they arise, must only exist for a limited period of time, namely, the time taken for investors and analysts to process new information and determine the impact on asset prices. In a largely efficient market this window of opportunity will be short yet the investor who can process information ahead of the pack may still profit from his knowledge.

Finally, it is worth pondering the way in which markets transmit information. For example, there are without doubt market participants whose motivations are driven by factors such as liquidity needs rather than profit. It is therefore conceivable that a trade which in turn determines a market price is driven not by new salient information concerning the asset but simply by the need to liquidate stock. This trade becomes market information to the other participants. Could it be construed as a negative view on the stock and contribute to a downward move in price? Generally the only information available is that a trade happened and not why it occurred.

15.3 Testing efficiency

Having introduced our concepts of efficiency, we must ask ourselves to what extent any of them represent real world markets? Certainly, trading strategies that result in superior returns do not seem to persist. This does not necessarily tell us much: once a strategy is known to make money, it is exploited and the advantage disappears. Note that this behaviour is similar to that underpinning no-arbitrage, although here it is not an arbitrage as such which is being eliminated. However, if enough strategies have been exploited then efficiency ought to approximately hold, since all these strategies are utilising most of the information.

Most tests of efficiency tend to focus on the third concept of weak efficiency which has the advantage of posing a more concrete problem. Put simply, one might attempt the following:

- take many time series;
- perform a statistical analysis;
- see what it tells us.

By contrast, testing semi-strong efficiency is problematic since to agree a test one has to agree on the amount of risk being taken and how much that risk should be discounted. One ends up having to assume CAPM, APT or some other equilibrium model to phrase the test. The test is then a test of the joint hypothesis that the model holds and that markets are efficient, and this is not much of a test.

15.4 Anomalies

We next consider certain anomalies in asset pricing and what these might in turn tell us about market efficiency. Importantly, we recognise that anomalies only disappear through their exploitation in pursuit of gain and hence we would not expect anomalies that do not allow the making of excess returns after transaction costs to disappear even in a highly-efficient market.

In this section, we discuss some known anomalies which have been used to counter market efficiency in debates to date.

1. *January effect.* Stocks make excess returns in January. This is particularly the case for small stocks, average return of 8% in January over 1941–1981. This contradicts efficiency. There are explanations in terms of tax effects and market microstructure. In countries with a January to December tax year, one expects some effects from people trying to minimise capital gains tax. However, the effect also exists in Australia which has a July to June tax year. It should disappear but does not seem to.
2. *May to October effect* There is a well-known phrase in the City of London being "sell in May and go away." Markets tend to be flat or go down in the (northern) summer.
3. *Non-zero auto-correlation.* Studies of stock returns have measured statistically significant non-zero auto-correlations. They are however tiny: e.g. 0.03.
4. *The size effect.* Over 1936–1977, small firms made about 20% more a year than large firms, where small firms were defined as the smallest fifth of the NYSE. Some possible explanations are
   - it simply reflects risk and pinning down risk is hard,
   - transactions costs are bigger,
   - it was true but is less true now.
5. *Rebound effect.* Suppose we compare two portfolios, one which invests in the 50 stocks that have lost the most in the previous five years and the second which invests in 50 stocks that have made the most in the previous five years. DeBondt and Thaler [4] found that the first portfolio has abnormally high returns and the second has abnormally low returns. This suggests that the market over-reacts to news. A crowd mentality causes the stock to move

too far. An alternative explanation is simply that it reflects additional risk in stocks that did badly in the last five years.

6. *Star fund managers.* Whilst most fund managers do not outperform the market, there are a few that seem to do so. This cannot be the case if markets are truly efficient. One example of such a fund manager is Jayesh Manek who shot to fame after winning the Sunday Times amateur investor competition two years running (1994 and 1995). This consisted of readers nominating stock portfolios and seeing whose makes the most at the end of the year. Naively, we would expect the probability of the same person winning twice in a row to be 1/(the number of entrants.) (Note no square here since someone has to win.) He made multiple entries so the probability is higher than that. In addition, the strategy for winning an investment competition is to take as much risk as possible rather than to invest sensibly. If you have taken substantially more risk than everyone else, you will have a greater chance of winning even if markets are perfectly efficient. However, a check on his fund's performance on 28/04/2006 showed that a five-year investor would have lost 21% percent of their investment as opposed to an investor in the FTSE all-share who would have made around 5%. Another check on 1/8/2012, found that a five-year investor would have lost 41% versus a gain of 1.6% for a FTSE investor. Note, however, his fund is a growth fund so it should do very well in good times and very badly in bad times. The last ten years have not been that good!

   Another interesting case is that of Warren Buffett, arguably the most famous fund manager of all investing through his vehicle Berkshire Hathaway and one of the world's richest people. In fifteen years Berkshire Hathaway's share rose from $5,500 to $90,000 largely reflecting the value of its holdings. Buffett does not appear to take excessive risk, so can we explain such returns? Possible explanations are:

   - Markets are more efficient now than they were 40 years ago.
   - Buffett doesn't just invest, he also sits on boards of companies and dabbles in various businesses such as reinsurance. Market efficiency does not say he cannot be a better manager.
   - Luck. The luck argument runs as follows: Fifty percent of investors in a given year will out-perform the median investor. Over a twenty-year period, one in a million will out-perform the median every year. Over thirty years, one in a billion. Hence there should be half a dozen star investors. In addition, Buffett did not outperform every year. Of course, this neglects the fact that most people living on the planet do not actively trade stocks. But do note that if one million people did, you would expect

one person to outperform the market every year for a twenty-year period by luck.

7. *Hedge funds.* The original hedge funds were investment companies that took a mixture of long and short positions in stocks. This made them neutral to the overall level of the market. They made money according to whether their views on stocks were correct. Some hedge funds have been very successful. On the other hand, The half-life of a hedge fund is pretty short. That a few do well is not so surprising and the ones that do badly disappear pretty quickly.

We conclude by commenting briefly on the problem of distinguishing genuine outperformance from investment where the risk profile allows for long periods of apparent outperformance, albeit with a possibility of catastrophic loss at some stage. This is an idea promoted heavily by Taleb, [25], in his book *Fooled by Randomness,* and the sequel *The Black Swan*, [26]. Remember that in the world of birds, black swans are rare and white swans are common. By analogy, suppose that you invest in such a way that one in X years you lose everything, but in other years you make a high return. That is, there is a small probability of dramatic loss. Until the small probability event, the "black swan," arrives and you lose everything, you appear very successful. If many investors pursue such strategies, then a few will survive for long periods of time. These few will appear to be geniuses and the rest will be forgotten. This is the concept of *survivorship bias* and it tells us that some apparent violations of efficiency may be no more than the adoption of an investment strategy carrying low probability of catastrophic loss. Apparent surplus returns over a period of years may simply reflect this implicit risk.

15.5 Conclusions on efficiency

What then do we conclude as to the efficiency of markets? Weak efficiency is supported to a large extent, but with imperfections as evidenced by very small but statistically significant non-zero auto-correlations. The semi-strong form seems to be impossible to test without a joint assumption as to an asset pricing model. Accordingly, this fails the principle of falsification. Strong form efficiency, whilst believed by a few, is more widely considered to be false.

15.6 Rationality

Rationality is closely related to efficiency. Rationality states that assets are accurately priced as the discounted value of future cash flows adjusted for risk. The risk adjustment can be made either adjusting the size of cash-flows as in CAPM or by using a risky discount curve.

That is, if a bond has a low credit rating then the total of its cashflows (coupons and principal) discounted using the risk-free rate will be higher than its value. A standard approach is to introduce the concept of a risky discount curve for each credit rating. The discount factors are lower than the corresponding factors from the risk-free discount curve. Accordingly, a risky discount curve expresses two things:

- the expected variation in size of the cash flow;
- the risk premium demanded by investors for the amount of variation.

Rationality is a stronger statement than efficiency and is therefore less likely to be true. To see this, recall that in our earlier discussion we noted that in an efficient market an asset price must reflect all information, but there is nothing to prevent additional fluctuations, so long as these are truly random in occurrence and hence do not allow for some cunning exploitation for profit. In a rational market this cannot be the case. Changes in asset price result solely from a change in the cashflow discounting calculation, i.e. changes in the timing and size of cashflows or the perception of risk associated with these cashflows. We examine here a few examples of market behaviours which argue against rationality.

1. *Excess Volatility.* Stocks move up and down much more than one would expect from the volume of information received. This suggests that people's risk preferences waver considerably. However, it is difficult to truly assess changes in risk preferences.
2. *Market Crashes.* Market crashes do occur. The FTSE lost 30% of its value in one day in October 1987. Interestingly, there was not much news that day or immediately before to account for this. Indeed, most of the big crashes of history have been news-less. Why then should risk premia or expectations have changed so drastically?
3. *Bubbles.* Markets from time to time enter bubbles. The value keeps on increasing to unimagined heights before bursting spectacularly. Examples of these include:
    - the South Sea Bubble;
    - tulip mania;

- the internet bubble;
- the Florida land bubble;
- the Great Crash of 1929.

The key characteristic of a bubble is that investors lose sight of fundamental value. Everyone else is making money and hence they do not wish to lose out. They invest despite a belief that the fundamentals are wrong, relying upon a general belief that the price will keep going up for a while longer and that they will exit in time. People certainly do make money in a bubble environment. Others, less fortunate, lose heavily when the crash happens.

15.7 Famous bubbles

By way of illustration, we pause to highlight three of the most notable bubble phenomena across history, the South Sea bubble, so-called Tulip mania and more recently the internet bubble.

The South Sea Company was a joint stock company formed in the UK in 1711. An early example of a public–private partnership, it was conceived as a way to restructure a 9 million pound national debt. All debt-holders would be issued with shares in the South Sea Company in exchange for government debt. To make the offer attractive, the company was granted a monopoly for trading between the UK and South America by the British government. Its share price went up by a factor of ten within a year as it expanded its operations dealing in government debt. However, with Britain involved at that time in a war against Spain, the dominant colonial power in South America, the company never achieved any real profits from its monopoly. After peaking in 1720, the share price collapsed to little above its flotation price with a fall of about 87%. When asked about the rising of the South Sea Company Stock, Sir Isaac Newton (who is believed to have lost 20,000 pounds (2 million dollars in today's terms)) is reputed to have answered "I can calculate the movement of the stars, but *not* the madness of men." Many other joint stock companies were formed at the same time and rode the bubble. In reaction to this crisis, it was made illegal to create a joint stock company without a Royal Charter: the "Bubble act" of 1720 which was not repealed until 1825. Some believe that it temporarily stifled capitalism in the UK. It has, however, been argued that the real motivation for the Bubble act was to keep the South Sea bubble going, [7], and, in fact, it became law some months before the bubble burst.

Another well-known example from European history is the Tulip mania, probably the most famous bubble of all time, which took place in 17th Century

Netherlands. In 1623, a tulip bulb cost 1,000 florins, six times the average annual wage. In 1635, 40 bulbs were sold for 100,000 florins. At one point, the price of a single bulb reached 6,000 florins. Fortunes were lost and made. There are always justifications that one can find for such a phenomenon. For example, it has been argued that at that time tulips had special properties caused by a virus – the mosaic virus; and that prices were driven up by the Thirty Years War of 1618–1648. Additionally, it has been argued, [27], that the crash was a result of a contractual conversion of futures prices to option exercise prices, as an attempt by Dutch buyers and officials to bail themselves out of speculative losses.

More recently, we have the internet bubble. As the internet took off in the late 1990s as new medium for communication and commerce, companies became very valuable purely for having ".com" in their names. The Nasdaq, where most such companies were listed, soared to immense heights before rapidly declining in 2000. Driving this increase in values was the widely-held belief that in the future all commerce would be performed via the internet and therefore all internet companies would make a fortune. Many people declared it to be a bubble at the time. To support market rationality, new stock valuation methods were invented to justify these prices although it suffices to say that these methods now enjoy less popularity.

We might reflect on an example in Shiller's book *Irrational Exuberance* [20] which was written in 1999 before the bubble burst. The internet business "eToys" launched in 1997 and went public in 1999 with a market capitalisation of around \$8 billion. The performance figures for this pure internet company provide a striking contrast to this market valuation as summarised by

- Total sales in 1998: \$30,000,000;
- Profits in 1998: $-\$28,600,000$.

We can contrast this with the familiar "Toys R Us" which had a market capitalisation at the same time of around \$6 billion and a large number of physical stores:

- Sales: \$11.2 billion;
- Profits: \$376,000,000.

The verdict of history has seen eToys go bankrupt whilst Toys R Us remains a familiar household name.

15.8 Justifying high stock prices

Over long periods of time, such as 30 years, stocks have outperformed other investments consistently in the US and UK. Therefore we might conclude that they are not risky if invested in for the long term. Consequently, they should not carry a large risk premium and prices should be much higher. In fact, it is a well-known problem in finance that stock prices appear to return much more than one would expect for the level of risk and is sometimes referred to as *the equity risk premium puzzle.*

However, the US has only had four or five periods of 30 years since its economy got going. On the other hand, the Japanese stock market is much lower than 20 years ago. Plenty of other countries have had stocks do rather dismally but they are small economies now, so no one notices. Hence we are faced with the problem of survivorship bias, again.

15.9 Further reading

Although it dates back to 1841, Mackay's *Extraordinary Popular Delusions and The Madness of Crowds*, [10], is still a good account of historical bubbles and it shows there's nothing new in them! For a rousing defense of market efficiency, see Ross's *Neoclassical Finance*, [18]. Ross himself recommends Shleifer's *Inefficient Markets: An Introduction to Behavioral Finance* for the counter arguments.

For convincing critiques of rationality, see the books by Smithers [24] and Shiller [20].

15.10 Review

By the end of the chapter the reader should be able to do the following.

1. State the three forms of stock market efficiency.
2. State which forms of market efficiency would be contradicted by a study showing that insider trading is an effective way to beat the market.
3. Describe what is meant by market rationality.
4. Describe the size, rebound and January effects.
5. Explain what is the equity risk premium puzzle.
6. Explain why excess volatility is an argument against market rationality.
7. Explain why crashes constitute an argument against market rationality.
8. Name three famous market bubbles.
9. Explain what is survivorship bias.

15.11 Questions

Question 15.1 If winning an investment competition is purely by chance and the same N participants enter each year, what is the probability that the same person wins it k years in a row?

Question 15.2 Suppose that markets are strongly efficient, and $10,000$ people enter an investment competition. The aim is to divide up a $1,000,000 between listed market companies, and the person with the biggest winnings after a year gets $10,000, while everyone else gets zero. A radio station gets their parrot to pick ten stocks at random and divides the money equally between them. Can you do better than the parrot and if so what strategy would you adopt, and why, to maximise the chances of winning?

Question 15.3 A bright honours student develops an algorithmic trading strategy for the buying and selling of gold. The strategy uses programmed trading with the inputs being the movements of stock prices and oil prices. He presents you with data showing that the strategy would have made a return of 1000% over the last year. Discuss what this would say about market efficiency and state how you would evaluate the evidence.

Question 15.4 Stockmarket rules require all share transactions by directors in companies to be reported to the stockmarket within 3 days. A study examines whether copying the directors' actions immediately after their trade leads to excess returns, and concludes that it does. A second study examines what would happen if the trading was done one day after the announcement, and concludes that it does not lead to excess returns. What would this say about the various forms of market efficiency?

16

Brownian motion and stock price models across time

16.1 Introduction

Thus far, we have focussed on models for stock prices at a single time horizon or on a discrete set of times. However, the hedging of complex products takes place in continuous time and models for the continuous movements of stock prices are essential for the pricing of derivatives products. In addition, when modelling liabilities and assets simultaneously for the hedging of complex products in the long term, it is necessary to study how quantities evolve through time.

16.2 Brownian motion

First, we develop the basics of the mathematics for modelling the log of the stock price via the most fundamental of such models: Brownian motion. It is the archetypal continuous stochastic process. Its importance is vast in a variety of fields ranging from physics to finance. The principal innovation from what we have studied so far is that we look at all time frames simultaneously. It is the joint distribution across many time frames that matters.

Definition 16.1 A process W_t on an interval $[0,T]$ is said to be a *Brownian motion* if:

- $W_0 = 0$;
- $W_t - W_s$ is normal of mean zero and variance $t-s$ for any s and t with $0 \leq s < t \leq T$;
- the increment $W_t - W_s$ is independent of the path of W_r for $r \leq s$;
- W_t has continuous sample paths, i.e. the map

$$t \mapsto W_t$$

is continuous.

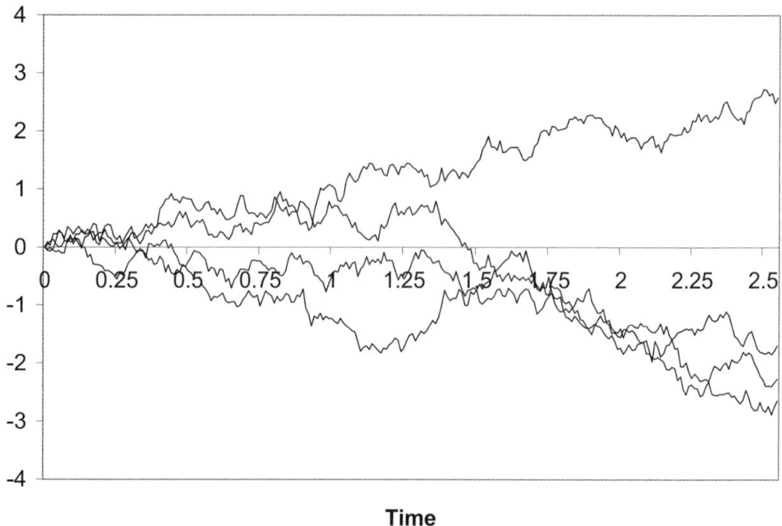

Figure 16.1 Paths of a Brownian motion

It is important to realise that two processes can have the same distribution at each time frame but be quite different. We give an example of a process that is a not Brownian motion. Suppose we let Z be normally distributed, $N(0,1)$, and take

$$X_t = \sqrt{t}Z.$$

Then X_t is $N(0,t)$ for all t but the increments are not independent.

It is not obvious that Brownian motion exists in a pure mathematical sense. However, it is possible to put a measure on the space of continuous paths that constructs Brownian motion. This is called Wiener measure. There are many ways to construct this measure. We illustrate some Brownian paths in Figure 16.1.

16.3 Differentiability properties of Brownian motion

Brownian motion has many strange properties. Typically, these hold *almost surely*; that is, with probability 1. We now develop a few of these relating to its differentiability. In fact, the following hold almost surely:

- paths are nowhere differentiable;
- paths are not increasing (or decreasing) on any given interval.

We prove only the latter. Consider a particular interval $[a,b] \subseteq [0,T]$ with $a < b$. Take an increasing sequence of points q_n inside $[a,b]$; for example, set

$$q_n = b - (b-a)/n.$$

If we can show that the probability that W_t is increasing on the set $\{q_n\}$ is zero then that will suffice, since the event is smaller than the event that W_t is increasing on $[a,b]$.

Let E_n denote the event that W_t is increasing on the first n points; that is, on the points q_1 to q_n. So E_n is the event

$$W_{q_1} \leq W_{q_2} \leq \cdots \leq W_{q_n}$$

and this is equivalent to

$$W_{q_j} - W_{q_{j-1}} \geq 0$$

for $j = 2, \ldots, n$, and these are independent and of probability 0.5. So E_n is of probability 2^{1-n}.

Now let E be the intersection of E_n over all n. The event E is that W_t is increasing on $\{q_n\}$ for all n. Since

$$E \subset E_n, \qquad \text{for all } n,$$

we have

$$\mathbb{P}(E) \leq \mathbb{P}(E_n) = 2^{1-n}$$

for all n and so is zero. The event that the path is increasing on $[a,b]$ is contained in E and so has probability zero.

Proving results about differentiability is a little hard for the level of this book but we prove something related. We can show that paths are not *Lipschitz continuous* anywhere.

Definition 16.2 A function f is *Lipschitz continuous* at s with constant K if there exists $\delta > 0$ such that

$$|t-s| < \delta \implies |f(t) - f(s)| \leq K|t-s|.$$

Note that Lipschitz continuity is actually much stronger than continuity. For example, the function f, defined by

$$f(t) = |t|^{\alpha},$$

is continuous at $t = 0$ for $\alpha \in (0,1)$, but is not Lipschitz continuous there, since

$$f(t)/|t| = |t|^{\alpha-1}$$

goes to infinity as t goes to zero.

Theorem 16.3 *For a given L the probability that the Brownian motion W_t on $[0,T]$ has a point of Lipschitz continuity with constant L is zero.*

Proof Rescaling, it is enough to consider the case that $T = 1$. Let E_n be the event that for *some* $s \in [0,1]$

$$|W_t - W_s| \leq L|t-s| \quad \text{whenever } |t-s| \leq 3/n.$$

Let

$$V_{k,n} = |W_{k/n} - W_{(k-1)/n}|$$

for $k = 1,\ldots,n$.

If E_n holds then we can get bounds on the size of some of the $V_{k,n}$ by considering the value of k such that

$$\frac{k-1}{n} < s \leq \frac{k}{n}.$$

If $k < 2$, take $k = 2$. If $k > n-1$, take $k = n-1$. Note that for any j

$$|W_{j/n} - W_{(j-1)/n}| \leq |W_{j/n} - W_s| + |W_s - W_{(j-1)/n}|,$$

and so taking $j = k-1, k, k+1$, we have

$$E_n \subseteq \bigcup_{k=2}^{n-1} \{V_{j,n} \leq 6Ln^{-1} \quad \text{for } j = k-1, k, k+1\}.$$

We then have

$$\begin{aligned}
\mathbb{P}(E_n) \leq & (n-2)(\mathbb{P}(V_{1,n} \leq 6Ln^{-1}))^3, \\
\leq & c(n-2)n^{-3/2}, \\
\leq & c' n^{-1/2},
\end{aligned}$$

for some constants c and c'. The events E_n are increasing but the bounds are going to zero. So the probability of E_n is zero all n. The probability of their union must therefore be zero, too, and the result follows. □

The reason that Lipschitz continuity fails is that the increments are not small enough as a function of t. We can make them smaller by squaring them. We can therefore examine Brownian motion's *quadratic variation.* By definition,

$$\mathbb{E}((W_t - W_s)^2) = t - s.$$

If we divide into $[0,T]$ into 2^k pieces then we can define

$$Q_k = \sum_{l=0}^{2^k-1} (W_{T2^{-k}(l+1)} - W_{T2^{-k}l})^2.$$

The expectation of Q_k is T. Its variance is

$$2^k \operatorname{var}(W_{T2^{-k}}^2).$$

For any s,

$$\begin{aligned}\operatorname{var}(W_s^2) =& \mathbb{E}((W_s^2 - s)^2),\\ =&\mathbb{E}(W_s^4 - 2W_s^2 s + s^2),\\ =&3s^2 - 2s^2 + s^2,\\ =&2s^2.\end{aligned}$$

So the variance of Q_k is

$$2T^2 2^{-k}.$$

This will go to zero as k goes to infinity. In fact, one can prove that Q_k converges to T almost surely. This says that the *dyadic quadratic variation* of W_t on the interval $[0,T]$ is T. Note that the general definition of quadratic variation requires one to consider all possible partitions rather than just the evenly-spaced ones which is why this is the dyadic case rather than the general one.

What does all this tell us? It means that Brownian motion is very jagged. Any differentiable function will have zero quadratic variation. For it to be positive, a function must be very rapidly moving up and down. Brownian motion is very far from differentiable.

16.4 Computing with Brownian motion

We do not work with Wiener measure directly. Instead, we compute using Brownian motion's properties.

Example 16.4 If W_t is a Brownian motion and

$$X_t = X_0 + at + \sigma W_t,$$

what is the probability that $X_t > y$?

It is

$$\mathbb{P}(\sigma W_t > y - X_0 - at) = \mathbb{P}(W_t > (y - X_0 - at)/\sigma).$$

Since W_t is normal with mean 0 and variance t this is

$$N(-(y - x_0 - at)/(\sigma\sqrt{t})).$$ ◇

Example 16.5 What is the probability that W_t is negative at time 1 and positive at time 2 if W_t is a standard Brownian motion?

We write

$$W_1 = X, \qquad W_2 = X + Y,$$

with X, Y independent standard normals.

The event is then

$$X < 0, \qquad Y > -X,$$

which can be rewritten as

$$X < 0, \qquad Y > 0, \qquad |Y| > |X|.$$

These three events are independent and of probability 0.5, so the answer is 1/8. ◇

16.5 More properties

We now look at some more properties of Brownian motion. We can consider the covariance between its values at two different times. We have

$$\mathrm{Cov}(W_s, W_t) = \min(s,t).$$

To prove this, we take $s \leq t$, and compute

$$\begin{aligned}
\mathbb{E}(W_s W_t) &= \mathbb{E}(W_s W_s + W_s(W_t - W_s)), \\
&= \mathbb{E}(W_s W_s) + \mathbb{E}(W_s(W_t - W_s)), \\
&= s + \mathbb{E}(W_s)\mathbb{E}(W_t - W_s), \quad \text{since independent}, \\
&= s = \min(s,t).
\end{aligned}$$

Note the general technique: decompose into a part that is determined at time s and a part that is independent of behaviour up to time s.

Another important property of Brownian motion is that it is Markovian; that is, its future distribution depends only on its current value: how it reached its current value is irrelevant. We can see this by writing

$$W_t = W_s + (W_t - W_s)$$

for $s < t$. We have that $W_t - W_s$ is independent of behaviour before time s so the only information required for future behaviour is W_s.

From one Brownian motion, we can construct others.

Proposition 16.6 *If W_t is a Brownian motion then so are:*

- $C_t = \frac{1}{\sqrt{c}} W_{ct}$;
- $X_t = -W_t$;
- $Y_t = W_{t+r} - W_r$ *for any* $r > 0$.

For C_t, it is obvious that increments have zero mean and are independent and normally distributed. The only thing that needs to be checked is that $C_t - C_s$ has the correct variance, however, this follows from an easy direct computation. For X_t and Y_t, all the properties are clear.

We can use the fact that C_t is a Brownian motion to deduce some curious properties. In particular, we can show the following.

Proposition 16.7 *The supremum and infimum of W_t on the interval $[0, \infty)$ are ∞ and $-\infty$ with probability* 1.

Proof The result for the infimum follows from that of the supremum using the fact that $-W_t$ is a Brownian motion. Consider the distribution of $\sup W_t$. We can regard it as a random variable taking values in the interval $[0, \infty]$. Since C_t is a Brownian motion for any c, its supremum has the same distribution as that of W_t. This means that the distribution of the supremum is invariant under scaling by c. It must therefore be supported in the two points 0 and ∞.

We now need to show that the probability of the supremum being zero is zero. Call the supremum Z. We have

$$\mathbb{P}(Z = 0) \leq \mathbb{P}(W_1 \leq 0 \quad \text{and} \quad W_s \leq 0, \quad \text{for all } s \geq 1).$$

Since Z has the same distribution as $\sup(W_{1+r} - W_1)$, the latter can only take values in 0 and ∞. So the events

$$\{W_1 \leq 0, \quad \text{and} \quad W_s \leq 0, \quad \text{for all } s \geq 1\}$$

and

$$\{W_1 \leq 0, \quad \sup(W_{1+r} - W_1) = 0, \quad \text{for all } r \geq 0\}$$

are the same.

We therefore have

$$\mathbb{P}(Z = 0) \leq \mathbb{P}(W_1 \leq 0 \quad \text{and} \quad \sup(W_{1+r} - W_1) = 0 \quad \text{for } r \geq 0).$$

The second term is just $\mathbb{P}(Z = 0)$, and it is independent of the event $W_1 \leq 0$. So

$$\mathbb{P}(Z = 0) \leq 0.5 \mathbb{P}(Z = 0)$$

and we have that $\mathbb{P}(Z = 0) = 0$ as required. □

We can use this result to prove the recurrence of Brownian motion; that is, it returns to every level with probability 1. Since we can make the argument at any time, it follows that it actually returns to every level infinitely often. Observe that for every j

$$W_{t+j} - W_j$$

is a Brownian motion starting at zero. It has a supremum of $+\infty$ and an infimum of $-\infty$. Since it is continuous, it must therefore visit every value after time j.

16.6 Arithmetic and geometric Brownian motions

A process W_t on an interval $[0,T]$ is said to be an *arithmetic Brownian motion* if:

- there exist μ, σ such that $W_t - W_s$ is normal of mean $\mu(t-s)$ and variance $\sigma(t-s)$ for any $0 \leq s < t \leq T$;
- the increment $W_t - W_s$ is independent of the path of W_r for $r \leq s$;
- W_t has continuous sample paths; that is, the map

  $$t \mapsto W_t$$

  is continuous.

In other words,

$$W_t = W_0 + \mu t + \sigma \tilde{W}_t,$$

where $\tilde{W}_t$ is a standard Brownian motion.

Arithmetic Brownian motion is not a great model for stock prices since it allows stock prices to go negative. It is also generally believed that stock prices change more in an absolute sense when prices are high. A stock will move more from day to day when it is $10,000$ than when it is 1. An arithmetic Brownian motion model says that stock prices changes are absolute and stock price increments do not depend on level, however.

For that reason, *geometric Brownian motion* is now more popular. We let the log stock price (all logs are base e) follow an arithmetic Brownian motion. So

$$\log S_t = \log S_0 + at + \sigma W_t,$$

with W_t a standard Brownian motion. Equivalently, we can write

$$S_t = S_0 e^{at+\sigma W_t}.$$

See Figure 16.2 for some typical paths.

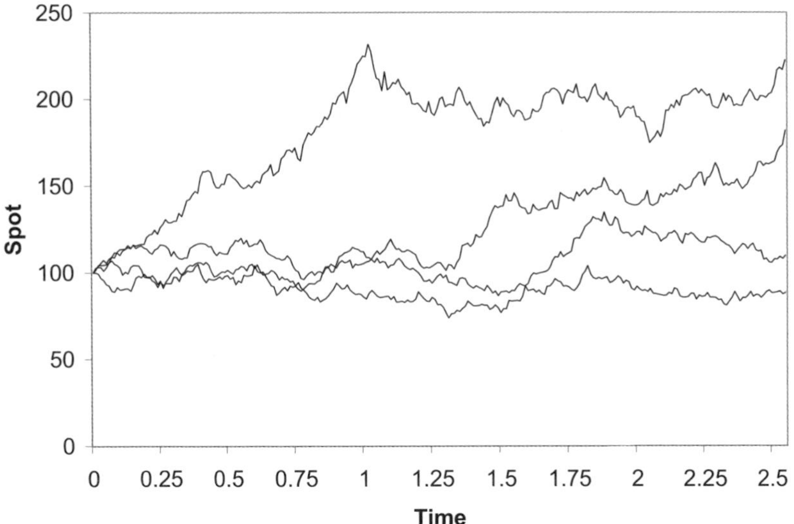

Figure 16.2 Paths of a geometric Brownian motion

By construction, S_t is always positive so we do not have to worry about negativity. Since the log stock price increments have a distribution independent of stock value, the stock price increments have a distribution which is proportional to the stock price, and so the model captures the idea that moves should be bigger when the stock price is high.

Clearly, S_t is log-normal and we shall see below that

$$\mathbb{E}(S_t) = S_0 e^{at+\frac{1}{2}\sigma^2 t}.$$

Because of this, we typically write

$$S_t = S_0 e^{(\mu - 0.5\sigma^2)t + \sigma W_t}.$$

We then have

$$\mathbb{E}(S_t) = S_0 e^{\mu t}.$$

We call μ the *drift* of the stock and σ the *volatility*.

Since the distribution of the stock is log-normal for each time horizon, using Brownian motion does not add much when we wish to study portfolio properties for single time horizons. However, when we wish to study the effects of continuous-time hedge rebalancing, it is an important tool. We refer the reader to [8] for more information.

16.7 Log-normal models for stock prices

Closely related to the geometric Brownian motion for stock price movements is the log-normal model. The difference essentially is that of viewpoint: when when we restrict to a single time frame or a small number of time frames we talk of log-normal models; when we are working in continuous-time we speak of geometric Brownian motion.

We have the following.

Theorem 16.8 *If*

$$\log S_t = \log S_0 + \mu t + \sigma\sqrt{t}Z$$

with Z a standard normal, then

$$\mathbb{E}(S_t) = S_0 e^{\mu t + \sigma^2 t/2},$$

and

$$\operatorname{Var} S_t = S_0^2 e^{(2\mu+\sigma^2)t}\left(e^{\sigma^2 t} - 1\right).$$

The first of these can be proved via direct integration against the normal density. The second follows from it and the fact that

$$\log S_t^2 = 2\log S_t,$$

and so if S_t is log-normal then so is S_t^2.

The quantity

$$\nu = \mu + \sigma^2/2,$$

is the *drift of the stock*. If $\sigma = 0$, then we have a riskless asset; in this case we take $\mu = r$ to be the continuously-compounding interest rate.

For $\sigma > 0$, we expect $\mu > r$ to compensate us for taking the risk. We define

$$\lambda = \frac{\nu - r}{\sigma}$$

to be the *market price of risk* for this stock. It is the compensation in terms of extra return for each unit of risk taken. Note different stocks can (and will) have different market prices of risk. Note also that this definition does not take into account any difference between diversifiable and undiversifiable risk.

The simple form of a log-normal random variable makes it easy to compute VAR. Suppose S_t is log-normal with parameters μ and σ. What is the VAR at level p?

We know

$$S_t = S_0 e^{\mu t + \sigma\sqrt{t}Z},$$

which is monotone increasing as a function of Z.

This means that the VAR at level p is simply found by:

- finding the VAR for Z;
- plugging it into this formula to get S_t at the VAR level;
- subtracting from S_0 to get the losses.

So we find that the VAR is

$$S_0 - S_0 e^{\mu t + \sigma\sqrt{t}N^{-1}(p)}.$$

Many criticisms have been made of the log-normal model for stock prices. The principal of these are:

- log returns are not normal, particularly over short time periods;
- volatility is stochastic with the market going through periods of high and low volatility;
- volatility tends to go up when stock prices go down;
- crashes occur.

One use of the log-normal model is to assess whether an investment is large enough to have a high probability of covering a liability. Thus if we have a liability, A, at t we can compute the probability that a log-normally distributed portfolio value, S_t, will exceed it.

If liability is A, then we need to find

$$\mathbb{P}(S_t \geq A) = \mathbb{P}(\log S_0 + \mu t + \sigma\sqrt{t}Z \geq \log A).$$

The right-hand side is equal to

$$\mathbb{P}\left(Z \geq \frac{\log A - \log S_0 - \mu t}{\sigma\sqrt{t}}\right)$$

or

$$N\left(-\frac{\log A - \log S_0 - \mu t}{\sigma\sqrt{t}}\right).$$

Note that all we can do is assess probabilities. If the probability is too low, we will need to invest more or adopt a different investment strategy. If we wished to be absolutely sure of covering the liability then we would need to invest in riskless assets.

In the very long term, drift wins. So provided $\nu > 0$, the log-normal model will predict that stock prices may become arbitrarily big if we wait long enough. Even if $\nu = 0$ the probability of being close to the starting value becomes very small as time gets large.

Whilst this is not necessarily a big problem for stock market indices, it is unrealistic for other sorts of financial quantities. e.g. volatility, inflation, interest rates, where we believe that there is a stable level which the quantity will tend to live near.

When using log-normal models, we have to be careful to distinguish between mean and median. Suppose we decide to model a liability at time t by a fixed amount A. Suppose we pick an investment in a stock index so that the mean is A, and use a log-normal model.

We estimate μ and σ and take an initial amount S_0 such that

$$S_0 e^{\mu t + 0.5\sigma^2 t} = A.$$

What then is the probability that $S_t > A$?

It is equal to

$$\mathbb{P}(\log S_0 + \sigma\sqrt{t}Z + \mu t > \log S_0 + \mu t + 0.5\sigma^2 t),$$

with Z standard $N(0,1)$, and equals

$$\mathbb{P}(Z > 0.5\sigma\sqrt{t}).$$

This goes to zero as $t \to \infty$! The mean and median are quite different in the long term for the log-normal distribution.

16.8 Auto-regressive processes

In practice, actuaries tend to use more complex processes when modelling long-term investments and liabilities in order to achieve greater realism. One building block is *auto-regressive processes.* These are also known as *mean-reverting*. The idea is that there is a natural value for a quantity and it will tend to drift back towards it over time.

We work with a discrete time model, namely, AR(1):

$$Y_t = \mu + a(Y_{t-1} - \mu) + \sigma\varepsilon_t.$$

The term μ is the long-term average. The term a is the speed of reversion to the mean, and σ is the volatility. The terms ε_t are independent $N(0,1)$.

We might consider

$$\log(S_{t+1}/S_t) = Y_t;$$

then Y_t describes the changes in $\log S_t$.

The AR(1) model captures the idea that the size of the increments mean-reverts but it does not produce *volatility clustering*; that is, the phenomenon that markets go into states of heightened volatility for periods of time.

One extension is therefore the ARCH(1) model; here CH stands for *conditional heteroskedastic*. We set

$$Y_t = \mu + a(Y_{t-1} - \mu) + \sigma_t \varepsilon_t,$$
$$\sigma_t^2 = a_0 + \alpha(Y_{t-1} - \mu)^2.$$

Note that the further Y_{t-1} is from μ, the bigger σ is (for $\alpha > 0$.)

We can generalise ARCH to get GARCH. Now we set

$$Y_t = \mu + a(Y_{t-1} - \mu) + \sigma_t \varepsilon_t,$$
$$\sigma_t^2 = a_0 + \alpha_1(Y_{t-1} - \mu)^2 + \beta \sigma_{t-1}^2.$$

Note that the further Y_{t-1} is from μ, the bigger σ is (for $\alpha > 0$.) The difference between GARCH and ARCH is that the value of σ_{t-1} has a role.

16.9 The Wilkie model

The Wilkie model was invented to address the following issues:

- life offices/super funds invest heavily in shares which have variable returns;
- fixed interest rates change over long periods of time;
- very long term liabilities must be covered;
- inflation affects liabilities, e.g. index-linked pensions.

The author's objective was to present a minimal model for the description of the total investments of a life assurance company. We therefore want a model for *long term* investment returns together with the economic indicators that affect the liabilities.

The modelled quantities are:

- $Q(t)$: an index of retail prices, (e.g. the consumer price index, CPI);
- $D(t)$: an index of share dividends;
- $Y(t)$: the dividend yield on the share index;
- $C(t)$: the consol yield; that is, the yield on bonds of infinite maturity.

The consol yield is so called because of a set of perpetual bonds issued by the British government known as consols. These pay a regular coupon and could in theory be redeemed. However, the coupon is sufficiently low that this is unlikely to occur and so they can be regarded as being of infinite maturity. Note that D and Y together determine the share price index, $P(t)$, since

$$Y(t) = D(t)/P(t).$$

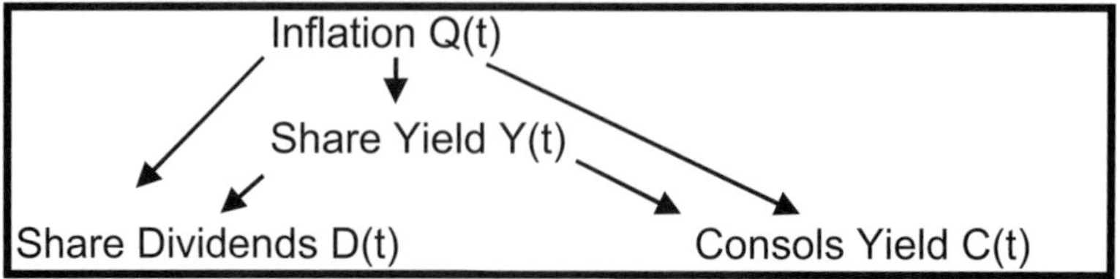

Figure 16.3 The quantities evolved in the Wilkie model and how they inter-relate

The quantity t is time in years and the model applies to year-on-year estimates.

The evolved quantities and how they relate to each other are illustrated in Figure 16.3. Inflation is the biggest driving factor: it affects the three other series. The share yield is deemed to have an effect on share dividends and fixed interest returns. The cascade structure makes calibration easier.

Modelling retail prices

It is the changes of

$$I(t) = \log Q(t) - \log Q(t-1)$$

that are being modelled. We are therefore really modelling the inflation rate rather than the index itself. We set

$$I(t) = \mu_I + a_I(I(t-1) - \mu_I) + \sigma_I Z_t^I,$$

where Z_t^I are independent $N(0,1)$ variables, and where

- μ_I is the long term mean,
- a_I is the speed of reversion,
- σ_I is the volatility of the process.

In other words, we are using a first-order auto-regressive model for inflation.

Dividend yields

These are modelled as a spread to inflation with the spread following an AR(1) process.

$$\log Y(t) = b_y I(t) + Y_s(t),$$

where

$$Y_s(t) = \mu_Y + a_Y(Y_s(t-1) - \mu_Y) + \sigma_Y Z^Y(t),$$

and $Z^Y(t)$ is a sequence of independent $N(0,1)$ variables.

Note that in this model, the dividend yield depends directly on inflation but not the converse.

Share dividend index

This is again modelled via its log:

$$D(t) = D(t-1)e^{K(t)},$$

with

$$K(t) = I_D(t) + \mu_D(t) + \Phi_D \varepsilon_Y(t-1) + \gamma_D \varepsilon_D(t-1) + \varepsilon_D(t),$$

and

$$\begin{aligned} I_D(t) =& \alpha_D M_D(t) + \beta_D I(t), \\ M_D(t) =& \delta_D(t) I(t) + (1-\delta_D) M_D(t-1), \\ \varepsilon_Y(t-1) =& \sigma_Y Z^Y(t-1), \\ \varepsilon_D(t) =& \sigma_D Z^D(t). \end{aligned}$$

We have parameters $\mu_D, \alpha_D, \gamma_D, \Phi_D$ and σ_Y.

Consol yield

The yield on long bonds is split into two parts: the *real part,* $C_n(t)$, and the *inflation part,* $C_m(t)$. We thus write

$$C(t) = C_m(t) + C_n(t),$$

where

$$C_m(t) = d_C I(t) + (1-d_C) I(t-1),$$

and

$$\log C_n(t) = \mu_C + a_C(\mu_c - \log C_n(t-1)) + y_C(\sigma_y Z_Y(t) + \sigma_C Z_C(t)).$$

Note the appearance of Z_Y here; this yields correlation with the dividend index and dividend yield.

16.10 Using the Wilkie model

Once we have the processes, we still need to calibrate the model and run simulations. For calibration, Wilkie used the data used from 1919 to 1982 at annual intervals to estimate the process parameters.

One then inputs starting values to represent the current market. Once this has been done, many paths are simulated. These are done one-step at a time with standard normal random variables being inserted. Once one has done many paths, the joint distribution of the quantities at the end has been simulated.

For example, to see the probability, p, that an inflation-linked liability is met, we simply run the simulation many times, N, and compute the number k that it was met and set

$$p = k/N.$$

This is an example of *Monte Carlo* simulation.

Note that as well as the modelled quantities we also get other quantities as outputs:

1. the share price index via $P(t) = D(t)/Y(t)$;
2. a "rolled-up" share index, $\mathrm{PR}(t)$ calculated by assuming all dividends after tax are reinvested into shares.

The second is useful in that it summarises what an investment portfolio in shares would be worth.

How can we test the Wilkie model? We can run the model many times and see whether the fitted values are plausible. For example, over a time period e.g. 1, 5, 10, 20, 50 or 100 years, we could measure the growth in the inflation index, use this to compute annual inflation and then see whether the implied value is believable. For example, is the mean value around 5%? Are we happy with that? Is the variance of inflation in the long term small, i.e., is the probability of hyperinflation too large in the model? The Wilkie model seems to be satisfactory on these. We can do similar analysis for each of the other time series.

The Wilkie model has a lot of parameters. One reason for the cascade structure is to allow each of the sets of parameters to be looked at separately making the calibration easier. Models with lots of parameters are generally regarded as dangerous. You can get lots of plausible fits leading to differing conclusions. One can therefore test the model to see how robust the conclusions are versus the parameter choices. Does a small change in inputs, lead to a big change in outputs?

Whilst the Wilkie model is plausible, there is not a great deal of evidence that it does particularly well at predicting future distributions. However, given the long-term nature of the model, it is very hard to test its predictions.

16.11 Review

In this chapter, we have examined the elementary properties of Brownian motion and introduced stock-price models. You should be able to do the following questions and tasks.

1. Define Brownian motion.
2. Prove that in a given interval the probability that a path of a Brownian motion is increasing is zero.
3. Give an example of a process such that X_t is distributed as $N(0,t)$ but is not Brownian motion.
4. Show that the probability that a Brownian motion is Lipschitz continuous at any point is zero.
5. Show that the mean and the variance of the dyadic quadratic variation of a Brownian tend to t and zero respectively as the number of points tends to infinity.
6. Define geometric Brownian motion.
7. What advantages does geometric Brownian motion (GBM) have over ordinary Brownian motion as a stock price model?
8. Define the drift and volatility of a GBM.
9. What is the log-normal model for stock price returns?
10. What are the mean and variance of a log-normally distributed asset price?
11. What is the definition of the market price of risk?
12. What are the disadvantages of using a log-normal model for stock prices?
13. What is a mean-reverting process?
14. Define ARCH and GARCH processes.
15. What are the variables in the Wilkie model?
16. How do the variables in the Wilkie model depend on each other?
17. What is the Wilkie model used for?
18. Describe how the Wilkie model might be used to assess a long-term liability.
19. Order the methodologies of mean–variance analysis, utility theory, CAPM, geometric means and stochastic dominance, in terms of how strong the assumptions needed on the investor are, with weakest first.

16.12 Questions

Question 16.1 What is the probability that $W_2 > 2$?

Question 16.2 What is the probability that $W_1 > 1$ and $W_2 > W_1$?

Question 16.3 Let P_n denote a sequence of discrete subsets of $[a,b]$, such that $P_n \subset P_{n+1}$. Suppose $a,b \in P_n$. Let $|P_n|$ denote the size of the largest interval in $[a,b] - P_n$. If $|P_n| \to 0$, show that the expectation of the quadratic variation defined using P_n converges to $b-a$. Show that its variance goes to zero.

Question 16.4 Let W_t be a Brownian motion on $[0,\infty)$. Let

$$X_t = t^{\alpha} W_{t\beta}.$$

For what values, if any, of α and β does X_t define a Brownian motion?

Question 16.5 In all of the situations below, W_t is a standard Brownian motion.

1. What is the probability that $W_j > j$?
2. What is the probability that $W_1 > 0$ and $W_2 < 0$?
3. Compute $\mathbb{E}(W_t^k | W_s)$ for $s < t$ and $k = 2,3,4$.

Question 16.6 If W_t is a standard Brownian motion on $[0,T]$ show that $W_T - W_{T-t}$ is also a Brownian motion.

Question 16.7 What is $\mathrm{Cov}(W_s, W_t)$ for a general arithmetic Brownian motion?

Question 16.8 Suppose W_t is a standard Brownian motion. Let $X_t = W_t$ for $t \leq 1$. For $t > 1$, let

$$X_t = W_1 - (W_s - W_1).$$

Does X_t define a Brownian motion?

Question 16.9
A function, f, is Hölder continuous on $[0,T]$ if there exists C, γ such that

$$|f(s) - f(t)| \leq C|s-t|^{\gamma}.$$

If $\gamma > 0.5$ show that f has zero dyadic quadratic variation.

Question 16.10 Suppose S_t follows a geometric Brownian motion. Show that S_t^k also does, for $k \in \mathbb{N}$.

Question 16.11 For each of the following, compute the market price of risk.

1. A stock has drift 0.10 and volatility 0.1. The risk-free continuously-compounding rate is 0.05.
2. A stock has drift 0.15 and volatility 0.2. The risk-free continuously-compounding rate is 0.08.
3. A stock has drift 0.075 and volatility 0.2. The risk-free continuously-compounding rate is 0.08.
4. A stock has drift 0.0 and volatility 0.3. The risk-free continuously-compounding rate is 0.03.

Question 16.12 An asset price is log-normal with drift 0.1 and volatility 0.1. The initial price is 100. What are the mean and variance of its price after 1, 2, and 3 years?

Question 16.13 An asset price is log-normal with drift 0.1 and volatility 0.3. The initial price is 100. What is the probability is worth more than 90 after 1, 2, and 3 years?

Question 16.14 An asset price is log-normal with drift 0.05 and volatility 0.2. The initial price is 100. What is the probability is worth more than 110 after 1, 2, and 3 years?

Question 16.15 An asset price is log-normal with drift 0.1 and volatility 0.1. The initial price is 100. What is its ten-day VAR at the 0.1%, 1%, and 5% levels?

Question 16.16 An asset price is log-normal with drift 0.05 and volatility 0.3. The initial price is 100. What is its one-day VAR at the 0.1%, 1%, and 5% levels?

Question 16.17 An asset price is log-normal with drift 0.25 and volatility 0.8. The initial price is 100. What is its 30-day VAR at the 0.1%, 1%, and 5% levels?

Question 16.18 An asset is worth 100. Let S_t be its value at time t. If $S_t + 25$ is log-normally distributed with drift 0.1 and volatility 0.2, what is its ten-day VAR at the 0.1%, 1%, and 5% levels?

Question 16.19 An asset is worth 100. Let S_t be its value at time t. If $S_t + 10$ is log-normally distributed with drift 0.05 and volatility 0.25, what is the probability that it exceeds the value of 110 after each of 1, 2, and 3 years?

Appendix A
Matrix algebra

A.1 Definitions

Much of modern portfolio theory is applied linear algebra or matrix algebra. Whilst we expect the reader to have encountered matrix algebra before, we briefly summarise the key ideas and techniques to jog the reader's memory.

Definition A.1 A matrix is a rectangular table of numbers. It will have m rows and n columns. It is then said to be of dimensions (m,n).

We can write a matrix, A, as

$$A = (a_{ij}) = \begin{pmatrix} a_{11} & a_{12} & \cdots & a_{1n} \\ a_{21} & a_{22} & \cdots & a_{2n} \\ & & \vdots & \\ a_{m1} & a_{m2} & \cdots & a_{mn} \end{pmatrix}.$$

A vector, v, with n entries can be regarded as a matrix with n rows and 1 column. An (m,n)-matrix can be applied to an n-vector to produce an m-vector, $w = Av$, via

$$(Av)_i = \sum_{j=1}^{n} a_{ij} v_j.$$

This means that we can use matrices to express linear equations. The equation

$$Av = w,$$

is equivalent to the linear system of equations

$$a_{r1}v_1 + a_{r2}v_2 + a_{rn}v_n = w_r \quad \text{for } r = 1,2,\ldots,m.$$

An important feature of the application of matrices is linearity. If $\lambda, \mu \in \mathbb{R}$ and

$x, y \in \mathbb{R}^n$, then

$$A(\lambda x + \mu y) = \lambda Ax + \mu Ay.$$

More generally, we can multiply matrices $A = (a_{ij})$ and $B = (b_{jk})$ together, provided the number of columns of A equals the number of rows of B, via

$$(AB)_{ik} = \sum_{j=1}^{n} a_{ij} b_{jk}.$$

The matrix AB then has the same number of rows as A and the same number of columns as B. The application of a matrix to a vector can be seen as a special case.

Matrix multiplication is associative: given matrices A, B and C,

$$A(BC) = (AB)C.$$

In the case that C is a vector, v, we can write

$$(AB)v = A(Bv).$$

The square matrix with 1s on the diagonal and zeros elsewhere is called the identity matrix and we have

$$AI = A = AI,$$

for all matrices when these multiplications are defined.

We can also define the transpose of a matrix A via

$$A^{\mathrm{T}}_{ij} = A_{ji}.$$

A matrix is said to be symmetric if $A = A^{\mathrm{T}}$.

If we regard vectors as single-column matrices then their transposes are single-row matrices. This means that given an $n \times n$ matrix, C, and an n-vector, v, we can compute the 1×1 matrix

$$v^{\mathrm{T}} C v = \sum_{i,j} v_i C_{ij} v_j$$

which we shall regard as a number since it only has one entry.

A square matrix A, is said to be invertible if there exists a matrix B such that

$$AB = I \quad \text{or} \quad BA = I.$$

If one of these equations is satisfied then the other is automatically satisfied as well, and B is said to the inverse of A, usually denoted A^{-1}. Not all square matrices are invertible, however.

If we need to solve a linear system,

$$Ax = y,$$

and we have found A^{-1} then we can set $x = A^{-1}y$, since

$$Ax = A(A^{-1}y) = (AA^{-1})y = Iy = y.$$

A.2 Gaussian elimination

Since the solution of linear systems can be achieved by finding a matrix inverse, much effort has gone into finding algorithms for inverting matrices. In this section, we discuss one method for both solving matrix equations and inverting matrices: *Gaussian elimination*. In fact, whilst the algorithm can be used to find the inverse, it can also be used to solve systems without finding it. It is based on the fact that the solution of a set of simultaneous equations is not changed by the following operations:

- multiplying a row by a non-zero constant;
- switching two rows;
- adding a multiple of one row to another.

The algorithm is as follows:

1. switch rows to get $a_{11} \neq 0$;
2. multiply row 1 by a_{11}^{-1};
3. subtract a_{j1} times row 1 from row j for each $j > 1$.

We now have a 1 in the top left and zeros below it. Now apply the same algorithm with the second row and second column, and so on. At the end, the system has become the identity matrix times a vector x and so is straightforward to solve.

Example A.2 Solve

$$\begin{pmatrix} 2 & 1 & 1 \\ 1 & 2 & 1 \\ 1 & 1 & 2 \end{pmatrix} \begin{pmatrix} x \\ y \\ z \end{pmatrix} = \begin{pmatrix} 2 \\ 4 \\ 6 \end{pmatrix}.$$

After step 1, we get

$$\begin{pmatrix} 1 & 1/2 & 1/2 \\ 1 & 2 & 1 \\ 1 & 1 & 2 \end{pmatrix} \begin{pmatrix} x \\ y \\ z \end{pmatrix} = \begin{pmatrix} 1 \\ 4 \\ 6 \end{pmatrix}.$$

Subtracting down the column:

$$\begin{pmatrix} 1 & 1/2 & 1/2 \\ 0 & 3/2 & 1/2 \\ 0 & 1/2 & 3/2 \end{pmatrix} \begin{pmatrix} x \\ y \\ z \end{pmatrix} = \begin{pmatrix} 1 \\ 3 \\ 5 \end{pmatrix}$$

Working with the second column:

$$\begin{pmatrix} 1 & 1/2 & 1/2 \\ 0 & 1 & 1/3 \\ 0 & 1/2 & 3/2 \end{pmatrix} \begin{pmatrix} x \\ y \\ z \end{pmatrix} = \begin{pmatrix} 1 \\ 2 \\ 5 \end{pmatrix}$$

Subtracting from the top and bottom rows:

$$\begin{pmatrix} 1 & 0 & 1/3 \\ 0 & 1 & 1/3 \\ 0 & 0 & 4/3 \end{pmatrix} \begin{pmatrix} x \\ y \\ z \end{pmatrix} = \begin{pmatrix} 0 \\ 2 \\ 4 \end{pmatrix}$$

We now work with the final column:

$$\begin{pmatrix} 1 & 0 & 1/3 \\ 0 & 1 & 1/3 \\ 0 & 0 & 1 \end{pmatrix} \begin{pmatrix} x \\ y \\ z \end{pmatrix} = \begin{pmatrix} 0 \\ 2 \\ 3 \end{pmatrix}$$

to get the solution:

$$\begin{pmatrix} 1 & 0 & 0 \\ 0 & 1 & 0 \\ 0 & 0 & 1 \end{pmatrix} \begin{pmatrix} x \\ y \\ z \end{pmatrix} = \begin{pmatrix} -1 \\ 1 \\ 3 \end{pmatrix} \qquad \diamond$$

If we wished to find the inverse of the matrix, we would simply solve the problem for the three target vectors

$$\begin{pmatrix} 1 \\ 0 \\ 0 \end{pmatrix}, \quad \begin{pmatrix} 0 \\ 1 \\ 0 \end{pmatrix}, \quad \begin{pmatrix} 0 \\ 0 \\ 1 \end{pmatrix}.$$

The three solutions written as the three columns of a matrix then give the inverse.

It is, of course, rare in this modern age for anyone working in financial services to solve a matrix system by hand. However, it is not so rare in examination halls and so it must still be learned.

Appendix B
Solutions

Chapter 1

Solution to Q1.1. We have

Probabilities	Returns	Returns squared	for semi-variance
0.1	–3	9	13.69
0.1	–2	4	7.29
0.2	–1	1	2.89
0.2	0	0	0.49
0.2	1	1	0
0.1	2	4	0
0.1	10	100	0

We conclude

$$\begin{array}{rl} \text{mean} & 0.7 \\ \text{variance} & 11.61 \\ \text{semi-variance} & 2.774 \end{array} .$$

Solution to Q1.2. The derivative goes through the sum and we get

$$y_i.$$

Solution to Q1.3. Here we write out the sums

$$\langle Au, v\rangle = \sum_i \sum_j A_{ij} u_j v_i.$$

$$\langle u, A^t v\rangle = \sum_j u_j \sum_i A^t_{ji} v_i = \sum_j u_j \sum_i A_{ij} v_i.$$

$$u^t A^t v = \sum_i u_i \sum_j A_{ji} v_j.$$

Up to labels these are all the same.

Solution to Q1.4.

$$g(x) = \sum x_i A_{ij} x_j.$$

When we differentiate, we get

$$\sum_{i,j} \frac{\partial}{\partial x_k}(x_i A_{ij} x_j).$$

Clearly, the derivative of x_i with respect to x_k is zero unless $i = k$. This means we get

$$\sum_j A_{kj} x_j + \sum_i x_i A_{ik}.$$

Since A is symmetric, this equals

$$2\sum_j A_{kj} x_j.$$

Solution to Q1.5. The covariance matrix is

$$\begin{pmatrix} 144 & 60 & 48 \\ 60 & 100 & 40 \\ 48 & 40 & 64 \end{pmatrix}.$$

We compute $x^t C x$ to get the variance and take the square root to get the standard deviation. With $x = (1/3, 1/3, 1/3)$:

variance	67.11111111
s.d.	8.192137152

With $x = (0.5, 0.25, 0.25)$:

variance	78.25
s.d.	8.845903006

Solution to Q1.6. The covariance matrix is

$$\begin{pmatrix} 100 & 50 & 50 \\ 50 & 100 & 50 \\ 50 & 50 & 100 \end{pmatrix}.$$

We compute $x^t Cx$ to get the variance and take the square root to get the standard deviation. With $x = (1/3, 1/3, 1/3)$:

variance	66.66666667
s.d.	8.164965809

With $x = (0.5, 0.25, 0.25)$:

variance	68.75
s.d.	8.291561976

Solution to Q1.7. Given this information our best estimates of the average variance and covariance are

$\overline{\text{Var}}$	49
$\overline{\text{Cov}}$	9.8

We then use the formula to get

N	10	100
Var	13.72	10.192

Solution to Q1.8, The expected returns are

$\mathbb{E}(R_A)$	11
$\mathbb{E}(R_B)$	13

The variances and covariances are

Cov	1.75
$\text{Var}(R_A)$	3.25
$\text{Var}(R_B)$	2.75

It follows that the correlation is 0.5854.

Solution to Q1.9. If the assets are independent then they have zero covariance with each other. The variance of a large portfolio then goes to zero as N goes to infinity.

If the mean covariance is $x > 0$, then the variance for N large is

$$\frac{1}{N}\overline{\mathrm{Var}(R_i)} + \frac{N-1}{N}x,$$

which clearly converges to x as N goes to infinity.

Solution to Q1.10. The matrix C_θ is trivially square and symmetric. So the issue is when is it positive definite? If $\theta \in [0,1]$ then so is $1-\theta$, hence

$$x^t C_\theta x = (1-\theta)x^t C_0 x + \theta x^t C_1 x \geq 0,$$

for any x.

For θ outside this range, it will generally not be a covariance matrix since if, for example, $C_1 = 0$, then if we take $\theta > 1$, and $x = (1,0,\ldots,0)$, we get

$$(1-\theta)C_{0,11}$$

which will be negative since $C_{0,11}$ is the variance of the first asset. Similarly for $\theta > 1$.

Solution to Q1.11. The only additional constraint for correlation matrices over covariance matrices is that there must be a 1 on the diagonal which will certainly be true for $\theta \in [0,1]$.

Solution to Q1.12. We can write the asset return as $X = \mu + \sigma Z$ with Z a standard normal. We compute

$$\begin{aligned}\mathbb{E}((X-\mu)^2 I_{X<\mu}) =& \mathbb{E}(\sigma^2 Z^2 I_{Z<0}),\\ =& \sigma^2 \mathbb{E}(Z^2 I_{Z<0}).\end{aligned}$$

By symmetry, the final expectation is 0.5 and the answer is $\sigma^2/2$.

Solution to Q1.13. Let

$$S = \int_{-\infty}^{\mu} f(x)(x-\mu)^2 dx,$$

and

$$V = \int_{-\infty}^{\infty} f(x)(x-\mu)^2 dx.$$

We can rewrite these, putting $y = x - \mu$, as

$$S = \int_{-\infty}^{0} f(y+\mu)y^2 dy,$$

$$V = \int_{-\infty}^{\infty} f(y+\mu)y^2 dy.$$

Since f is symmetric about μ, we also have

$$S = \int_{0}^{\infty} f(y+\mu)y^2 dy.$$

We conclude $V = 2S$ for any density f with these properties, and so f and g will both give the ratio 2.

Solution to Q1.14. We use the linearity of expectation; if Z is a random variable and α is a constant then

$$\mathbb{E}(\alpha Z) = \alpha \mathbb{E}(Z).$$

So

$$\begin{aligned}
\operatorname{Cov}(\lambda X, \mu Y) =& \mathbb{E}((\lambda X - \mathbb{E}(\lambda X))(\mu X - \mathbb{E}(\mu X))), \\
=& \mathbb{E}(\lambda(X - \mathbb{E}(X))\mu(X - \mathbb{E}(X))), \\
=& \lambda\mu\mathbb{E}((X - \mathbb{E}(X))(Y - \mathbb{E}(Y))), \\
=& \lambda\mu \operatorname{Cov}(X, Y).
\end{aligned}$$

Solution to Q1.15. We compute

$$\begin{aligned}
\operatorname{Cov}\left(\sum_i x_i R_i, \sum_j y_j R_j\right) &= \mathbb{E}\left(\sum_i x_i (R_i - \overline{R}_i) \sum_j y_j (R_j - \overline{R}_j)\right), \\
&= \mathbb{E}\left(\sum_{i,j} x_i y_j (R_i - \overline{R}_i)(R_j - \overline{R}_j)\right), \\
&= \sum_{i,j} x_i y_j \mathbb{E}\left((R_i - \overline{R}_i)(R_j - \overline{R}_j)\right), \\
&= \sum_{i,j} x_i y_j \operatorname{Cov}(R_i, R_j), \\
&= x^t C y.
\end{aligned}$$

Solution to Q1.16. It is enough to consider the case where $\alpha = 1$. We can just multiply the random variables by $\sqrt{\alpha}$ to go from this to the general case. To go in the reverse direction, we divide by $\sqrt{\alpha}$.

If $\beta > 1$, we show that it is not a covariance matrix. The variance of a portfolio consisting of 1 unit of the first asset, -1 units of the second and zero of the rest, is

$$2 - 2\beta.$$

This is less than zero so $\beta \leq 1$ if it is a covariance matrix.

We now show that for $\beta \in [0,1]$, it is a covariance matrix. Take independent normal random variables Z_j for $j = 0, \dots, n$.

For $j > 0$, let

$$W_j = \sqrt{\beta} Z_0 + \sqrt{1-\beta} Z_j,$$

we then have that the variance of W_j is 1 and the covariance between W_i and W_j is β. So we have a constructed a set of random variables with this covariance matrix. So it is a valid covariance matrix.

For the general case, we simply need $\alpha \geq \beta$.

Chapter 2

Solution to Q2.1. We need to compute their means and variances. We get

(1) 2,3;
(2) $e^{0.5} = 1.649, e^2 - e = 8.64 - 2.71 = 5.73$;
(3) $2, 8/6 = 1.33$;
(4) $3, 8/6 = 1.33$;
(5) $5/2, 25/12$.

In order of means we have

$$(4) > (5) > (3) > (1) > (2).$$

In order of variances, we have

$$(3) = (4) < (5) < (1) < (2).$$

(4) has the highest mean and lowest (equal) variance so is better than all the others.

(2) has the lowest mean and highest variance, so is worse than all the others.

(5) and (3) are not comparable but both are better than (1).

So $(4) > (5), (3) > (1) > (2)$.

Solution to Q2.2. The expected returns are $5, 6, 7$ and the variances are $13, 6, 9$. The first asset is therefore less efficient than the other two. No more can be said since asset three has both higher variance and expected return than asset two.

Solution to Q2.3. We are given

$$\begin{array}{cc} \overline{R}_A & 10 \\ \overline{R}_B & 5 \\ \sigma_A & 8 \\ \sigma_B & 6 \\ \rho & 0 \end{array}$$

We use the formula for the minimal variance portfolio to get

$$\begin{array}{cc} X_A* & 0.36 \\ X_B* & 0.64 \end{array}$$

We then compute again to get

$$\begin{array}{cc} \overline{R}_P & 6.8 \\ \mathrm{Var}(P) & 23.04 \\ \sigma_P & 4.8 \end{array}$$

To maximise return without short-selling, put everything in A.

Solution to Q2.4. We divide into four parts, (a), (b), (c) and (d), according to the correlation.

(a) We have

$$\begin{array}{cc} \overline{R}_C & 5 \\ \overline{R}_D & 10 \\ \sigma_C & 10 \\ \sigma_D & 5 \\ \rho & -1 \end{array}$$

We compute

$$\begin{array}{cc} \mathrm{Var}(C) & 100 \\ \mathrm{Var}(D) & 25 \\ \mathrm{Cov}_{CD} & -50 \end{array}$$

Using the formula for the minimal variance portfolio, we get

X_C^*	0.333333333
X_D^*	0.666666667

We compute

$\overline{R}_P$	8.333333333
$\text{Var}\,P$	0
σ_P	0

We repeat for (b) to get

$\overline{R}_C$	5
$\overline{R}_D$	10
σ_C	10
σ_D	5
ρ	0
$\text{Var}(C)$	100
$\text{Var}(D)$	25
Cov_{CD}	0
X_C^*	0.2
X_D^*	0.8
$\overline{R}_P$	9
$\text{Var}\,P$	20
σ_P	4.47

We repeat for (c) to get

$\overline{R}_C$	5
$\overline{R}_D$	10
σ_C	10
σ_D	5
ρ	0.5
$\text{Var}(C)$	100
$\text{Var}(D)$	25
Cov_{CD}	25
X_C^*	0
X_D^*	1
$\overline{R}_P$	10
$\text{Var}\,P$	25
σ_P	5

We repeat for (d) to get

$\overline{R}_C$	5
$\overline{R}_D$	10
σ_C	10
σ_D	5
ρ	1
$\text{Var}(C)$	100
$\text{Var}(D)$	25
Cov_{CD}	50
X_C^*	–1
X_D^*	2
$\overline{R}_P$	15
$\text{Var}\,P$	0
σ_P	0

In all cases, the expected maximal return without short-selling is achieved by putting everything into D to get 10.

Solution to Q2.5. With correlation -1 and 1 we obtain a a V-shape. With correlation 0, it is a non-degenerate hyperbola.

Solution to Q2.6. For correlation -1, the opportunity set is a V shape, and the no-short-selling part is at the tip of the V. Note that the no-short-selling part will always contain the tip of the V since the formula for its weights give a ratio of two positive numbers. So the no short-selling efficient frontier will be a straight line going up from the tip, which has zero standard deviation.

For correlation 0, the opportunity set is a U shape, and the no-short-selling part is at the tip of the U. Note again that the no-short-selling part will always contain the tip of the V since the formula for its weights give a ratio of two positive numbers. So the no-short-selling efficient frontier will be the part of a U shape on the upper part of the hyperbola containing the tip of the hyperbola.

For correlation 1, the opportunity set is a V shape. The minimal variance portfolio will always involve short selling since we have to balance one asset against the other. If assets are S and C we have that the weight of C for minimal variance (which is zero) is

$$\frac{\sigma_S^2 - \sigma_S\sigma_C}{\sigma_C^2 + \sigma_S^2 - 2\sigma_S\sigma_C}.$$

The denominator is $(\sigma_S - \sigma_C)^2$ so will not be negative.

The numerator is $\sigma_S(\sigma_S - \sigma_C)$, so the weight of C will be positive if and only if it has smaller standard deviation. Similarly for S.

We therefore have two possibilities according to the expected returns of S and C. Either the no-short-selling opportunity set lies in the upper part of the V and is also efficient; or it lies in the bottom part of the V, and in that case, only its left-most point is efficient, and we get a single point.

Note that if S and C are perfectly correlated and have the same standard deviation then they are essentially the same asset except for level of return, and all money would go into the one with higher expected return. If they have the same expected return as well, then they are effectively the same asset!

Solution to Q2.7. Stocks have higher return and lower variance so are preferable in a mean–variance sense.

For equal amounts:

$$\begin{array}{rc} X_C^* & 0.5 \\ X_D^* & 0.5 \\ \mathrm{Var}(P_E) & 332.25 \\ \sigma_E & 18.23 \\ \overline{R}_E & 11 \end{array}$$

We do the usual computations to get the minimal variance portfolio statistics:

$$\begin{array}{rc} \mathrm{Var}(O) & 625 \\ \mathrm{Var}(S) & 484 \\ \mathrm{Cov}_{OS} & 110 \\ X_O^* & 0.421 \\ X_S^* & 0.579 \\ \overline{R}_P & 11.159 \\ \mathrm{Var}(P) & 326.66 \\ \sigma_P & 18.07 \end{array}$$

The market manipulation would increase the oil price when stocks are high so it would increase correlation and increase the variance of portfolios which hold a positive amount of each.

Chapter 3

Solution to Q3.1. We compute

$$\frac{\overline{R}_A - R_f}{\sigma_A}$$

in each case to get

R_f	$\overline{R}_A$	σ_A	λ
8	12	10	0.4
0	1	1	1
5	15	5	2
8	6	3	-0.667

The last case is clearly very strange.

Solution to Q3.2. We compute

$$X_A = \frac{\sigma_B}{\sigma_A}$$

and

$$X_A\overline{R}_A + (1 - X_A)R_f$$

in each case, to get

X_A	X_f	$\overline{R}_B$
2	–1	18
2	–1	15
0.2	0.8	2.8

Solution to Q3.3. We compute

$$X_A = \frac{\overline{R}_B - R_f}{\overline{R}_A - R_f}$$

and the standard deviation is

$$X_A\sigma_A$$

in each case, and we get

X_A	X_f	σ_B
0.333	0.667	3.33
2	-1	10
0	1	0

Solution to Q3.4. We find the efficient portfolio with the minimal expected return, and see if that has sufficiently low standard deviation. Since the portfolios lie on a straight line in expected return/standard deviation space, if this portfolio has too high a standard deviation, so will all portfolios with sufficiently high expected return.

We compute

$$X_A = \frac{\overline{R}_B - R_f}{\overline{R}_A - R_f}$$

and the standard deviation is

$$X_A \sigma_A$$

in each case, and we get

X_A	X_f	σ_B
0.5	0.5	8
1.714	-0.714	8.57
2.667	-1.667	18.667

So in all these cases the desired standard deviation is achievable.

Solution to Q3.5. To get the tangent portfolio, we divide the weights of the risky assets by their sum, which in this case is 0.8. We get

$$0.25,\ 0.3125,\ 0.1875,\ 0.25.$$

The expected return of the tangent portfolio is

$$5 \times 0.25 + 5 \times 0.3125 + 6 \times 0.1875 + 7 \times 0.25 = 5.4375.$$

To get a return of 10, we compute first that we get an excess return over the risk-free rate of 2.4375 for each unit of tangent portfolio.

So we need

$$\frac{10-3}{2.4375} = 2.871795$$

units of tangent portfolio.

The weights are therefore

X_A	0.7179487
X_B	0.897435897
X_C	0.538461538
X_D	0.717948718
X_f	−1.871794872

Solution to Q3.6. The covariance matrix is

$$\begin{pmatrix} 25 & 5 & 5 & 5 \\ 5 & 25 & 5 & 5 \\ 5 & 5 & 25 & 5 \\ 5 & 5 & 5 & 25 \end{pmatrix}.$$

We compute that the standard deviation of the tangent portfolio is $10/3$. So to get a standard deviation of 1, we take $3/10$ of the tangent portfolio, and $7/10$ of the riskless asset. For 10 we get 3 units of tangent portfolio. So we get

Target s.d.	1	10
Weight of tangent	0.3	3
Weight of risk free	0.7	–2
X_A	0.075	0.75
X_B	0.025	0.25
X_C	0.075	0.75
X_D	0.125	1.25

Solution to Q3.7. The covariance matrix is

$$\begin{pmatrix} 25 & 0 & 0 & 0 \\ 0 & 25 & 0 & 0 \\ 0 & 0 & 25 & 0 \\ 0 & 0 & 0 & 25 \end{pmatrix}.$$

We compute that the standard deviation of the tangent portfolio is 2.766. The best return will be achieved by holding a mixture of the tangent portfolio and the riskless asset. To get a standard deviation of 5, we therefore take 1.807 units of the tangent portfolio, and -0.807 of the riskless asset.

Target s.d.	5
Weight of tangent	1.807392228
Weight of risk free	–0.807392228
X_A	0.25819889
X_B	0.25819889
X_C	0.516397779
X_D	0.774596669

Solution to Q3.8. Our input data is

R_{f_1}	1
R_{f_2}	5
$\overline{R}_A$	10
$\overline{R}_B$	12
σ_A	11
σ_B	13
ρ	0
σ_1	8.439
$\overline{R}_1$	10.933
σ_2	8.516
$\overline{R}_2$	11.001

We first compute variances and covariances.

$\text{Var}(A)$	121
$\text{Cov}(A,B)$	0
$\text{Var}(B)$	169

We use the usual formula for the minimal variance portfolio with two assets to get

X_{MV}	0.5828
$\overline{R}_{MV}$	10.8345
σ_{MV}	10.78

Since expected return is linear in each asset's return, we solve a linear equation in the expected returns to get the portfolio weights

X_1	0.5335
Y_1	0.4665
X_2	0.4995
Y_2	0.5005

For finding the efficient portfolios with given expected returns. We have to identify whether its long risk-free asset, no risk-free asset or shorting risk-free asset.

There are two ways to do this problem. The first is to see where the expected return lies. If it is below the expected return of the first tangent portfolio then it can be achieved by lending and holding the first tangent portfolio. If it is between the expected returns of the tangent portfolios, then it is achievable by holding just the two tangent portfolios. If it is above the tangent portfolio for borrowing's expected return then we borrow and hold the tangent portfolio for borrowing.

The second way is to first try using the risk-free asset for lending. If that works we are done. If it says we should borrow then it does not work, so we try borrowing. If that says we should borrow then we are done. If it says we should lend, and the lending rate says we should borrow then we should do neither. In that case, we use a mix of the two tangent portfolios.

We work through the problem using the second way. We first try going long the risk-free asset and use the first tangent portfolio. The risk-free rate is 1.

We solve a linear equation in the expected returns of the risk-free asset and the tangent portfolio to find the weights.

We get

Return	Weight of tangent	Weight of R.F.	X	Y	σ
3	0.2013	0.7987	0.1074	0.0939	1.6992
6	0.5034	0.4966	0.2685	0.2348	4.2480
10.95	1.0017	−0.0017	0.5344	0.4673	8.4534

With 10.95, we see that borrowing is required and this will be true for any higher expected return so there is no point in computing with 12. The first two cases are now done.

We now try the last two cases using borrowing of the risk-free asset so the riskless rate is 5. We get

Return	Weight of tangent	Weight of R.F.	X	Y	σ
10.95	0.9915	0.0085	0.4953	0.4962	8.4436
12	1.1665	−0.1665	0.5827	0.5838	9.9337

The 10.95 does not involve borrowing so we conclude that it is in the part of the efficient frontier that does not involve the risk-free asset.

The 12 does involve borrowing and it is now done.

We now find the linear combination of the two tangent portfolios that gives the right expected return.

Return	Weight of T1	Weight of T2	X	Y	σ
10.95	0.7500	0.2500	0.5250	0.4750	8.4547

This has the desired characteristics and we have done all four cases.

For finding the required standard deviations, we can proceed similarly using either of the two approaches. We again use the second. For the long risk-free asset part we have that the standard deviation is equal to the weight of the tangent portfolio times its standard deviation, so we simply divide to get the weight.

Target s.d.	Weight of T1	Weight of R.F.	X	Y	$\overline{R}$
3	0.3555	0.6445	0.1897	0.1658	4.5311
6	0.7110	0.2890	0.3793	0.3317	8.0622
8.47	1.0037	–0.0037	0.5355	0.4682	10.9695

As before the first two are correct but the third involves shorting the risk-free asset.

We try the third and fourth cases doing borrowing with rate 5.

Target s.d.	Weight of T2	Weight of R.F.	X	Y	$\overline{R}$
8.47	0.9946	0.0054	0.5306	0.4640	10.9010
12	1.4091	–0.4091	0.7518	0.6574	13.3603

The 12 case is now done since it involves borrowing.

For the 8.47 case, we have to find an efficient portfolio that does not involve the risk-free asset. This not surprising since its standard deviation lies between that of the two tangent portfolios.

It is easier to solve for the correct variance since the formula for the variance of two portfolios is a quadratic and does not involve square-roots.

We can either solve as a linear combination of the two tangent portfolios and then infer the linear combination of the original assets, or we can just solve for the linear combination of the two assets. The equation we have to solve, with X the holding of asset A, is

$$X^2 \operatorname{Var}(A) + 2X(1-X)\operatorname{Cov}(A,B) + (1-X)^2 \operatorname{Var}(B) = 8.47^2.$$

This is a quadratic in X and easily solved. We have to take care to use the solution which lies between the two tangent portfolios.

We obtain

$$X = 0.5178, Y = 0.4823,$$

and the expected return is 10.966.

Note that we could alternatively have predicted on which part of the efficient frontier the portfolio lies by looking at where its expected return or standard deviation was relative to those of the two tangent portfolios.

Solution to Q3.9. Our input data is

R_{f_1}	2
R_{f_2}	6
$\overline{R}_A$	9
$\overline{R}_B$	12
σ_A	12
σ_B	13
ρ	0.5
σ_1	11.3
$\overline{R}_1$	11.07
σ_2	12.61
$\overline{R}_2$	11.83

We first compute variances and covariances.

$\mathrm{Var}(A)$	144
$\mathrm{Cov}(A,B)$	78
$\mathrm{Var}(B)$	169

We use the usual formula for the minimal variance portfolio with two assets to get

X_{MV}	0.5796
$\overline{R}_{MV}$	10.26
σ_{MV}	8.8456

Since expected return is linear in each asset's return, we solve a linear equation in the expected returns to get the portfolio weights:

X_1	0.5335
Y_1	0.4665
X_2	0.4995
Y_2	0.5005

For finding the efficient portfolios with given expected returns, we have to identify whether it is long risk-free asset, no risk-free asset or shorting risk-free asset. We first try going long the risk-free asset and use the first tangent portfolio. The risk-free rate is 2.

We solve a linear equation in the expected returns of the risk-free asset and the tangent portfolio to find the weights.

We get

Return	Weight of T1	Weight of R.F.	X	Y	σ
3	0.1103	0.8897	0.0342	0.0761	1.2459
6	0.4410	0.5590	0.1367	0.3043	4.9835
11.5	1.0474	–0.0474	0.3247	0.7227	11.8357

With 11.5, we see that borrowing is required and this will be true for any higher expected return so there is no point in computing with 12. The first two cases are now done.

We now try the last two cases using borrowing of the risk-free asset so the riskless rate is 6. We get

Return	Weight of T2	Weight of R.F.	X	Y	σ
11.5	0.9434	0.0566	0.0535	0.8899	11.8962
12	1.0292	–0.0292	0.0583	0.9708	12.9777

The 11.5 does not involve borrowing so we conclude that it is in the part of the efficient frontier that does not involve the risk-free asset.

The 12 does involve borrowing and it is now done.

We now find the linear combination of the two tangent portfolios that gives the right expected return.

Return	Weight of T1	Weight of T2	X	Y	σ
11.5	0.4342	0.5658	0.1667	0.8333	11.9594

This has the desired characteristics and we have done all four cases.

For finding the required standard deviations, we proceed similarly. For the long risk-free asset part we know that the standard deviation is equal to the weight of the tangent portfolio times its standard deviation, so we simply divide to get the weight.

Target s.d.	Weight of T1	Weight of R.F.	X	Y	$\overline{R}$
3	0.2655	0.7345	0.0823	0.1832	4.5472
6	0.5310	0.4690	0.1646	0.3664	7.0944
12	1.0619	–0.0619	0.3292	0.7327	12.1887

As before the first two are correct but the third involves shorting the risk-free asset.

We try the third and fourth cases doing borrowing with rate 6.

Target s.d.	Weight of T2	Weight of R.F.	X	Y	$\overline{R}$
12	0.9516	0.0484	0.2950	0.6566	11.1303
15	1.1895	–0.1895	0.3688	0.8208	13.4128

The 15 case is now done since it involves borrowing.

For the 12 case, we have to find an efficient portfolio that does not involve the risk-free asset. This not surprising since its standard deviation lies between that of the two tangent portfolios.

It is easier to solve for the correct variance since the formula for the variance of two portfolios is a quadratic and does not involve square-roots.

We can either solve as a linear combination of the two tangent portfolios and then infer the linear combination of the original assets, or we can just solve for the linear combination of the two assets. The equation we have to solve, with X the holding of asset A, is

$$X^2 \operatorname{Var}(A) + 2X(1-X)\operatorname{Cov}(A,B) + (1-X)^2 \operatorname{Var}(B) = 12^2.$$

This is a quadratic in X and easily solved. We have to take care to use the solution which lies between the two tangent portfolios.

We obtain

$$X = 0.1592, Y = 0.8408$$

and the expected return is 11.709.

Solution to Q3.10. The crucial point to observe here is that the investors wishes to be "fully invested in two Australian stocks." This means he will not put money in the risk-free asset.

This means that the investments are two parts. Either we will put money in a linear combination of the two stocks, or he will put it in a multiple of the tangent portfolio and borrow.

If be borrows he will hold $1+X$ units of the tangent portfolio with 8% risk-free rate and $-X$ units of the risk-free asset. Since he will borrow and not lend, we must have $X \geq 0$. The geometry in weight space of this part is therefore a (half-) straight line of the form

$$(-X, (1+X)a_1, (1+X)a_2), \qquad X \geq 0,$$

for constants a_1, a_2 with $a_1 + a_2 = 1$, in weight space.

In return/standard deviation space for this part, we have a straight line starting at the tangent portfolio and going off to infinity.

For the other part, we have a hyperbola going from the minimal variance portfolio to the tangent portfolio in return/standard deviation space.

In weight space, we have a straight line describing positive linear combinations of the minimal variance portfolio and the tangent portfolio with no risk-free holding. Once we get to the tangent portfolio, the line changes direction and we get multiples of the tangent portfolio and negative amounts of the risk-free asset.

To find the possible weights we need to find a_1 and a_2. This could be done by plotting the hyperbola in return/standard deviation space for varying values of weights in the two assets. Find the tangent line through $(0, 8)$ to get the expected return and standard deviation of the tangent portfolio, and then solve for the weights which yield that expected return.

Solution to Q3.11. We have not been given the tangent portfolio for lending, but only for borrowing. However, note that if it is possible to construct a portfolio using a long position in the risk-free asset and the borrowing portfolio then that is sufficient, since there will be an even better portfolio.

We compute

$$X_A = \frac{\overline{R}_B - R_f}{\overline{R}_A - R_f}$$

and the standard deviation is

$$X_A \sigma_A.$$

For the first case, 0.5 units of each of A and the risk-free asset achieve the desired return and standard deviation so there must be some efficient portfolio that does at least as well on each so this is possible.

In the last two cases, a borrowing at the risk-free rate is indicated since expected return is higher than that of the borrowing tangent portfolio.

X_A	X_f	σ_B
1.833	–0.833	9.167
3.5	–2.5	24.5

It is therefore possible in all three cases.

Solution to Q3.12. We have not been given the tangent portfolio for lending, but only for borrowing. However, note that if it is possible to construct a portfolio using the risk-free asset and the borrowing portfolio then that is sufficient.

For the first case, 0.5 units of each of A and the risk-free asset achieve the desired return and standard deviation so there must be some efficient portfolio that does at least as well on each so this is possible.

For the second and third cases, the return is greater than that of B so we must borrow. We solve to get the portfolio with precisely the desired standard deviation.

X_A	X_f	σ_B
3.5	–2.5	17.5
–1.5	2.5	–10.5

In the last case, we see a negative amount of the tangent portfolio. This reflects the fact that the borrowing rate is higher than the expected return on the tangent portfolio so borrowing is pointless. The best we can do is to invest in the tangent portfolio and not borrow, so the desired return cannot be achieved using the tangent portfolio and borrowing. This does not however prove that there is not some portfolio in the market that does have the desired characteristics, so in truth there is insufficient data to answer the question.

In the second case, we have that the standard deviation is in the desired range and so it is possible.

Chapter 4

Solution to Q4.1.

$$\begin{pmatrix} 1 & 1 \\ 1 & 1 \end{pmatrix} \begin{pmatrix} 1 \\ -1 \end{pmatrix} = \begin{pmatrix} 0 \\ 0 \end{pmatrix}.$$

Solution to Q4.2. Our input data is

$\overline{R}_1$	11
$\overline{R}_2$	12
σ_1	10
σ_2	11
ρ	0
r_f	1

We compute the covariance matrix to be

$$C = \begin{pmatrix} 100 & 0 \\ 0 & 121 \end{pmatrix}.$$

The inverse of the covariance matrix is

$$C^{-1} = \begin{pmatrix} 0.01000 & 0.00000 \\ 0.00000 & 0.00826 \end{pmatrix}.$$

The value of $\tilde{R} = R - R_f e$ is

$$\begin{pmatrix} 10 \\ 11 \end{pmatrix}.$$

We apply C^{-1} to $\tilde{R}$ to get

$$\begin{pmatrix} 0.1 \\ 0.0909 \end{pmatrix}.$$

We then divide by the sum of the entries to obtain the tangent portfolio:

$$x = \begin{pmatrix} 0.5238 \\ 0.4762 \end{pmatrix}.$$

We compute $x^t Cx$ to get the variance of returns of the tangent portfolio and obtain

$$54.88.$$

Taking square roots the standard deviation is $\sigma = 7.408$. The return of the tangent portfolio is $\overline{R}_T = 11.48$.

We now have that the slope of the investment line is

$$\frac{\overline{R}_T - R_f}{\sigma} = 1.414.$$

The investment line is therefore

$$\overline{R}_P = 1 + 1.414\sigma.$$

To get the weights of the MV portfolio, we use

$$e = \begin{pmatrix} 1 \\ 1 \end{pmatrix}$$

instead of $\tilde{R}$ to obtain

$$\begin{pmatrix} 0.54751 \\ 0.45249 \end{pmatrix}.$$

Solution to Q4.3. Our input data is

$\overline{R}_1$	11
$\overline{R}_2$	12
σ_1	6
σ_2	7
ρ	0
r_f	5

We compute the covariance matrix to be

$$C = \begin{pmatrix} 36 & 0 \\ 0 & 49 \end{pmatrix}.$$

The inverse of this covariance matrix is

$$C^{-1} = \begin{pmatrix} 0.02778 & 0.00000 \\ 0.00000 & 0.02041 \end{pmatrix}.$$

The value of $\tilde{R} = R - R_f e$ is

$$\begin{pmatrix} 6 \\ 7 \end{pmatrix}.$$

We apply C^{-1} to $\tilde{R}$ to get

$$\begin{pmatrix} 0.16667 \\ 0.14286 \end{pmatrix}.$$

We then divide by the sum of the entries to get that the tangent portfolio is

$$x = \begin{pmatrix} 0.53846 \\ 0.46154 \end{pmatrix}.$$

We compute $x^t C x$ to get the variance of returns of the tangent portfolio and obtain

$$20.8757.$$

Taking square roots the standard deviation is $\sigma = 4.569$. The return of the tangent portfolio is $\overline{R}_T = 11.462$.

We now have that the slope of the investment line is

$$\frac{\overline{R}_T - R_f}{\sigma} = 1.414.$$

The investment line is therefore

$$\overline{R}_P = 5 + 1.414\sigma.$$

To get the weights of the MV portfolio, we use

$$e = \begin{pmatrix} 1 \\ 1 \end{pmatrix},$$

instead of $\tilde{R}$ and get

$$\begin{pmatrix} 0.57647 \\ 0.42353 \end{pmatrix}.$$

Solution to Q4.4. We have to solve

$$Cy = e,$$

where e is all ones. The solution can be found by Gaussian elimination (or matrix inversion) and turns out to be

$$\begin{pmatrix} 1 \\ 0 \\ 0 \end{pmatrix}.$$

Since the sum of these entries are one, this is the minimal variance portfolio and it has expected return 11 and variance 1.

For the tangent portfolio, we have to solve

$$Cy = \tilde{R} = \begin{pmatrix} 6 \\ 9 \\ 12 \end{pmatrix}.$$

The solution is

$$y = \begin{pmatrix} -1 \\ 1 \\ 6 \end{pmatrix}.$$

The sum of the entries of y is 6 so we get weights

$$x = \begin{pmatrix} -1/6 \\ 1/6 \\ 1 \end{pmatrix}.$$

Its expected return is 17.5. The tangent portfolio's standard deviation is

$$\sqrt{x^t Cx} = 1.443.$$

Solution to Q4.5. We solve $Cy = e$ using Gaussian elimination or matrix inversion, the solution is

$$\begin{pmatrix} 0.5 \\ 0 \\ 0.5 \end{pmatrix}.$$

The expected return is 10. The variance is given by

$$x^t Cx = 1.$$

Solution to Q4.6. The tangent portfolio is given by

$$v = \frac{1}{\langle y, e\rangle} C^{-1}x.$$

We need to compute

$$v^t C v.$$

This equals

$$\begin{aligned}\frac{1}{\langle y, e\rangle^2}(C^{-1}x)^t C(C^{-1}x) &= \frac{1}{\langle y, e\rangle^2} x^t (C^{-1})^t CC^{-1}x, \\ &= \frac{1}{\langle y, e\rangle^2} x^t C^{-1}x, \\ &= \frac{1}{\langle y, e\rangle} x^t v.\end{aligned}$$

Note that we have used the fact that C is symmetric which implies that C^{-1} is also symmetric. To see this take transposes of the equations

$$CC^{-1} = I = C^{-1}C,$$

and use the fact that inverses of square matrices are unique.

Solution to Q4.7. We divide top and bottom by x^2 to get

$$\frac{\alpha x^{-2} + \beta x^{-1} + \gamma}{\delta x^{-2} + \varepsilon x^{-1} + \nu}.$$

Letting x to infinity makes x^{-1} and x^{-2} go to zero. So the answer is

$$\gamma/\nu.$$

Solution to Q4.8. Here we use the Tobin separation theorem. The ratio of the amount of each asset held by B to the amount held by A should be the same for all risky assets. This means that we multiply each of A's holdings by $0.25/0.1 = 2.5$ to get B's holdings.

We get

$$0.25,\ 0.275,\ 0.3,\ 0.325,\ 0.35,\ 0.375.$$

The risk-free asset is chosen to make the holdings sum to 1 and we get -0.875.

Solution to Q4.9. Since he cannot borrow we will hold non-negative amounts of the risk-free asset and the tangent portfolio if he puts money on deposit. If he wants higher expected returns he will hold zero risk-free assets and varying amounts along the efficient frontier.

In weight space, we therefore have a straight line of the form

$$(Xv, 1-X) \qquad \text{for } X \leq 1,$$

where v is the weights of the tangent portfolio and $1-X$ is the holding of the risk-free asset. Once we reach $X=1$, then we get a change of direction and get

$$(Xv+(1-X)w, 0),$$

where w is the weights of another efficient portfolio: the line changes direction.

In return/standard deviation space we have a straight line from the risk-free asset to the tangent portfolio and then a hyperbola.

Solution to Q4.10. In general, this portfolio will not be efficient for this investor. The asset X_n may well have been hedging exposure to one of the assets and so there is no reason to believe that the new portfolio is efficient.

Solution to Q4.11. The variance of R_n is zero so this is the standard result that discarding the risk-free asset from an efficient portfolio results in an efficient portfolio.

Solution to Q4.12. The asset D is risk-free. So we treat it separately. First we find the tangent portfolio in A, B, and C. We have to solve

$$Cy = \tilde{R},$$

with

$$C = \begin{pmatrix} 4 & 3 & 3 \\ 3 & 4 & 3 \\ 3 & 3 & 3 \end{pmatrix}.$$

We have

$$\tilde{R} = \begin{pmatrix} 4 \\ 3 \\ 2 \end{pmatrix}.$$

We solve to get

$$y = \begin{pmatrix} 2 \\ 1 \\ -2.3333 \end{pmatrix}.$$

Dividing by the sum of the weights, we get

$$x = \begin{pmatrix} 3 \\ 1.5 \\ -3.5 \end{pmatrix}.$$

The tangent variance is given by

$$x^t C x = 14.25,$$

and the expected return is 11.5.

To get the target return we use 0.5 units of the tangent portfolio and 0.5 units of D.

Its standard deviation is then

$$0.5\sqrt{14.25} = 0.5 \times 3.77 = 1.88.$$

Solution to Q4.13. Since there is a risk-free asset and we have mean–variance investors, they all hold multiples of the tangent portfolio plus some holding of the risk-free asset.

We are given that X holds

$$\begin{pmatrix} 0 \\ -1 \\ 1 \\ 1.5 \\ 0.5 \\ -1 \end{pmatrix}$$

where the last element is the risk-free asset.

The tangent portfolio is found by summing the weights of the risky assets, and then dividing by the sum. The sum is 2. So the tangent portfolio is

$$\begin{pmatrix} 0 \\ -0.5 \\ 0.5 \\ 0.75 \\ 0.25 \end{pmatrix}.$$

The tangent portfolio has expected return 16, variance 3.75, and standard deviation 1.936.

To get expected return of 8.5, we use 0.5 units of the tangent portfolio and 0.5 of the risk-free asset. The holding for Y is therefore

$$\begin{pmatrix} 0 \\ -0.25 \\ 0.25 \\ 0.375 \\ 0.125 \\ 0.5 \end{pmatrix}.$$

To get Z's holding, we note that the target standard deviation is $\sqrt{30} = 5.477$. The amount of tangent portfolio is therefore

$$5.477/1.936 = 2.828.$$

We multiply the weights of the tangent portfolio by this and use the risk-free asset to balance the weights to one. The answer is

$$\begin{pmatrix} 0 \\ -1.414 \\ 1.414 \\ 2.121 \\ 0.7071 \\ -1.828 \end{pmatrix}.$$

Chapter 5

Solution to Q5.1. The variance of the residual risk is

$$\sigma_{e_i}^2/N$$

so the standard deviation is

$$\frac{\sigma_{e_i}}{\sqrt{N}}.$$

So we get $\sigma_{e_i}/\sqrt{10}$, $\sigma_{e_i}/\sqrt{20}$, and $\sigma_{e_i}/\sqrt{100} = \sigma_{e_i}/10$.

Solution to Q5.2. If t is the target amount we have to solve

$$\frac{20}{\sqrt{N}} = t,$$

so

$$N = \frac{400}{t^2}.$$

We therefore get 4, 400, and 40,000.

Solution to Q5.3. We use the formula

$$\alpha_i + \beta_i \mathbb{E}(R_m)$$

to get the expected returns and we get

$$\begin{array}{lr} \mathbb{E}(R_A) & 31 \\ \mathbb{E}(R_B) & 22 \\ \mathbb{E}(R_C) & 8 \end{array}$$

For the covariance matrix, we use the formula

$$\beta_i^2 \sigma_m^2 + \sigma_{e_i}^2$$

to get the variances, and

$$\beta_i \beta_j \sigma_m^2$$

to get the covariances. We obtain

$$\begin{pmatrix} 190 & 120 & 30 \\ 120 & 100 & 20 \\ 30 & 20 & 35 \end{pmatrix}.$$

Solution to Q5.4. The f_i do not affect expected return so as in a single factor model we get

$$\alpha_i + \beta_i \mathbb{E}(R_m).$$

The cross-expectation will not affect variances so we get as in a single-factor model

$$\beta_i^2 \sigma_m^2 + \varepsilon.$$

For the covariance, we get the additional term

$$\mathbb{E}(f_i f_j)$$

and so the answer is

$$\beta_i \beta_j \sigma_m^2 + \delta.$$

Solution to Q5.5. First, we have to find the vector of expected returns and covariance matrix. We use the single-factor model formulas to get

$$\begin{array}{lc} \mathbb{E}(R_A) & 6 \\ \mathbb{E}(R_B) & 12 \\ \mathbb{E}(R_C) & 22 \end{array}$$

and obtain the covariance matrix of

$$\begin{pmatrix} 2 & 2 & 4 \\ 2 & 5 & 8 \\ 4 & 8 & 17 \end{pmatrix}.$$

We then carry out the usual algorithm to find the tangent portfolio. We get

$$\tilde{R} = \begin{pmatrix} 3 \\ 9 \\ 19 \end{pmatrix},$$

$$y = \begin{pmatrix} -1.409090909 \\ 0.181818182 \\ 1.363636364 \end{pmatrix},$$

and so after dividing by the sum, we find that the weights are

$$x = \begin{pmatrix} -10.333 \\ 1.333 \\ 10 \end{pmatrix}.$$

The expected return is 174 and the standard deviation is 35.41.

Solution to Q5.6. First we have to find the vector of expected returns and covariance matrix. We use the single-factor model formulas to get

$$\begin{array}{lc} \mathbb{E}(R_A) & 34 \\ \mathbb{E}(R_B) & 18 \\ \mathbb{E}(R_C) & 10 \end{array}$$

and obtain the covariance matrix of

$$\begin{pmatrix} 49 & 24 & 12 \\ 24 & 13 & 6 \\ 12 & 6 & 4 \end{pmatrix}.$$

we then carry out the usual algorithm to find the tangent portfolio. We get

$$\tilde{R} = \begin{pmatrix} 32 \\ 16 \\ 8 \end{pmatrix},$$

$$y = \begin{pmatrix} 0.5 \\ 0.25 \\ 0.125 \end{pmatrix},$$

and so after dividing by the sum, we find that the weights are

$$x = \begin{pmatrix} 0.5714 \\ 0.2857 \\ 0.1429 \end{pmatrix}.$$

The expected return of the tangent portfolio is 26 and the standard deviation is 5.237.

We now need to compute how many units of the tangent portfolio to get a variance of 10, which means a standard deviation of 3.162.

We obtain

$$z = 3.1623/5.237 = 0.6038.$$

So we multiply the weights of the tangent portfolio by z to get

$$0.3450, \quad 0.1725, \quad 0.0863$$

and 0.3962 units of the risk-free asset.

The expected return is 16.49 and the standard deviation of returns is 3.1623.

Solution to Q5.7. The beta is the ratio of the covariance of the stock with the market divided by the variance of the market. The variance of the market is 25. The covariances are 5 and 15. The betas are therefore 0.2 and 0.6.

The alpha is the expected return minus the beta times the market's expected return. So we get 13 and 14.

Solution to Q5.8. We compute

$$0.8 \times 0.8 + 0.23 = 0.87,$$
$$0.8 \times 1.2 + 0.23 = 1.19.$$

Solution to Q5.9. We compute

$$0.9 \times 0.8 + 0.13 = 0.85,$$
$$0.9 \times 1 + 0.13 = 1.03,$$
$$0.9 \times 1.2 + 0.13 = 1.21.$$

Solution to Q5.10. A and B should have the properties that $A+B$ is close to 1 and that A is less than 1. This suggests that the first one is best since the sum is 0.984.

This implies betas of

$$0.8572, \quad 0.984, \quad 1.1108.$$

Chapter 6

Solution to Q6.1. For a full covariance matrix the number of covariance entries is

$$N(N+1)/2 = 15,$$

and we need 5 means so we get 20.

For a single-factor model, we have

$$3N + 2 = 17.$$

For a two-factor model, we have

$$2N + 2L + LN = 10 + 4 + 10 = 24.$$

So in this case, more data is required for the multi-factor model.

Solution to Q6.2. We plug into the formula for expected returns to get

$$9.6, \quad 13.9;$$

and for variances to get

$$19.49, \quad 40.18.$$

The covariance is 25.55.

Solution to Q6.3. We plug into the formula for expected returns to get

$$3.9, \quad 4.1;$$

and for variances to get

$$4.28, \quad 2.48.$$

The covariance is 3.16.

Solution to Q6.4. We plug into the formula for expected returns to get

$$14.4, \quad 10.9;$$

and for variances to get

$$69, \quad 34.89.$$

The covariance is 11.9.

Solution to Q6.5. We compute the covariance matrix using the formulas to get

$$\begin{pmatrix} 6 & 4 \\ 4 & 6 \end{pmatrix}.$$

The expected returns are

$$21, \quad 26.$$

The $\tilde{R}$ vector is

$$\begin{pmatrix} 20 \\ 25 \end{pmatrix}.$$

We solve as usual to get that the tangent portfolio is

$$2/9, \quad 7/9.$$

The minimal variance portfolio is found in the usual way:

$$0.5, \quad 0.5.$$

Solution to Q6.6. We compute the covariance matrix. This is a little complicated, since we have covariance between the two indices. However, we can compute the covariance of a linear combination of correlated random variables in the usual way and then add on the idiosyncratic term.

$$\begin{pmatrix} 7 & 5.25 \\ 5.25 & 7 \end{pmatrix}.$$

The expected returns are

$$21, \quad 26.$$

The $\tilde{R}$ vector is

$$\begin{pmatrix} 13 \\ 18 \end{pmatrix}.$$

We solve as usual to get that the tangent portfolio is

$$-\frac{2}{31}, 1\frac{2}{31}.$$

The minimal variance portfolio is found in the usual way:

$$0.5, \quad 0.5.$$

Solution to Q6.7. If we have W_j, $j = 0, 1, \ldots, n$, which are independent standard normal $N(0,1)$ random variables, we can put for $j > 0$

$$Z_j = \alpha W_0 + \sqrt{1-\alpha^2} W_j.$$

The correlation between Z_j and Z_k for $k \neq j$ is α^2. So putting

$$\alpha = \sqrt{\rho},$$

we are done.

Note that W_0 is our common factor and W_j for $j > 0$ are the idiosyncratic risk.

Solution to Q6.8. We can use Question 6.7. We can rewrite

$$e_i = \varepsilon_i P_i$$

with ε_i a positive constant and P_i a standard normal random variable. Replacing P_i by $\sqrt{\rho} W_0 + \sqrt{1-\rho} W_j$, we obtain random variables with the same distribution, but with one extra common factor and uncorrelated residuals. So we simply put

$$I_{l+1} = W_0.$$

Solution to Q6.9. We apply Gram–Schmidt. We set

$$Y_1 = X_1 = \begin{pmatrix} 1 \\ 0 \\ 1 \end{pmatrix}.$$

We compute

$$\langle X_1, X_2 \rangle = 4, \text{ and } \langle X_1, X_1 \rangle = 2.$$

We therefore subtract 2 units of X_1 from X_2 to get

$$Y_2 = \begin{pmatrix} -1 \\ 2 \\ 1 \end{pmatrix}.$$

We compute

$$\langle Y_2, Y_2 \rangle = 6, \langle Y_1, X_3 \rangle = 2, \text{ and } \langle Y_2, X_3 \rangle = 8.$$

So we subtract 1 unit of Y_1 from X_3 and $4/3$ units of Y_2 from X_3 to get

$$Y_3 = \begin{pmatrix} 4/3 \\ 4/3 \\ -4/3 \end{pmatrix}.$$

Solution to Q6.10. We apply Gram–Schmidt. We set

$$Y_1 = X_1 = \begin{pmatrix} 1 \\ 0 \\ 1 \end{pmatrix}.$$

We compute

$$\langle X_1, X_2 \rangle = 2, \text{ and } \langle X_1, X_1 \rangle = 2.$$

We therefore subtract 1 unit of X_1 from X_2 to get

$$Y_2 = \begin{pmatrix} 0 \\ 2 \\ 0 \end{pmatrix}.$$

We compute

$$\langle Y_2, Y_2 \rangle = 4, \langle Y_1, X_3 \rangle = 3, \text{ and } \langle Y_2, X_3 \rangle = 10.$$

So we subtract 1.5 of Y_1 from X_3 and 10 units of Y_2 from X_3 to get

$$Y_3 = \begin{pmatrix} -1/2 \\ 0 \\ 1/2 \end{pmatrix}.$$

Solution to Q6.11. We apply Gram–Schmidt. We set

$$Y_1 = X_1 = \begin{pmatrix} 1 \\ 1 \\ 1 \end{pmatrix}.$$

We compute

$$\langle X_1, X_2 \rangle = 4, \text{ and } \langle X_1, X_1 \rangle = 3.$$

We therefore subtract $4/3$ units of X_1 from X_2 to get

$$Y_2 = \begin{pmatrix} -1/3 \\ 2/3 \\ -1/3 \end{pmatrix}.$$

We compute

$$\langle Y_2, Y_2 \rangle = 2/3, \langle Y_1, X_3 \rangle = 1, \text{ and } \langle Y_2, X_3 \rangle = -1/3.$$

So we subtract $1/3$ units of Y_1 from X_3 and -0.5 units of Y_2 from X_3 to get

$$Y_3 = \begin{pmatrix} 0.5 \\ 0 \\ -0.5 \end{pmatrix}.$$

Solution to Q6.12. Nothing! The whole point is to express the **same** stock returns in a different way – if they were different then the procedure would have failed. When we replace an index I_j by a linear combination of uncorrelated indices J_l we do this so in such a way that all means variances and covariances are unchanged.

Solution to Q6.13. Yes! It is clearly symmetric and it follows from the basic properties of integration that

$$\int_0^1 (\lambda f(x) + \mu g(x))h(x)dx = \lambda \int_0^1 f(x)h(x)dx + \mu \int_0^1 g(x)h(x)dx$$

which says

$$\langle \lambda f + \mu g, h \rangle = \lambda \langle f, h \rangle + \mu \langle g, h \rangle.$$

This leaves us to show that

$$\langle f, f \rangle = 0$$

implies that $f = 0$. However, if we take a polynomial and square it we obtain a polynomial that is nowhere negative. If it is zero everywhere between 0 and 1 then it is zero everywhere since it has an infinite number of zeros; otherwise it is only zero at a finite number of points and so must have positive integral. So if $f \neq 0$ we have

$$\langle f, f \rangle \neq 0,$$

as required.

Solution to Q6.14. First note that

$$\int_0^1 x^k dx = \frac{1}{k+1}.$$

We now apply Gram–Schmidt to the functions $1, x, x^2, x^3$. So

$$g_0 = 1,$$

and we see

$$g_1 = x - \frac{1}{2}.$$

We have

$$\langle g_1, g_1 \rangle = \frac{1}{12}.$$

For g_2 we have

$$\langle x^2, g_0 \rangle = \langle x^2, 1 \rangle = \frac{1}{3},$$
$$\langle x^2, g_1 \rangle = \langle x^2, g_1 \rangle = \langle x^2, x - \frac{1}{2} \rangle = \frac{1}{4} - \frac{1}{6} = \frac{1}{12}.$$

We therefore take

$$g_2 = x^2 - \left(x - \frac{1}{12}\right) - \frac{1}{3} = x^2 - x + \frac{1}{6}.$$

Solution to Q6.15. First note

$$\int_{-1}^1 x^j dx = \begin{cases} \frac{2}{j+1} & \text{if } j \text{ even} \\ 0 & \text{if } j \text{ odd.} \end{cases}$$

We take $g_0 = 1$ and $g_1 = x$, since

$$\langle 1, x \rangle = 0.$$

We have

$$\langle 1,1\rangle = 2, \qquad \text{and} \qquad \langle x,x\rangle = \frac{2}{3}.$$

We find

$$\begin{aligned}\langle x^2,1\rangle =& \frac{2}{3},\\ \langle x^2,x\rangle =& 0.\end{aligned}$$

So we set

$$g_2 = x^2 - \frac{1}{3},$$

and we are done.

Chapter 7

Solution to Q7.1. It is simplest to compute second derivatives; we find

- $f'' = 2$;
- $f'' = 6x$;
- $f'' = -\frac{1}{4}(x+1)^{-1.5}$;
- $f'' = 0$.

All but the third are convex since they have non-negative f'' for $x \geq 0$. The last two are concave since they have non-positive f'' for $x \geq 0$.

Solution to Q7.2. It is simplest to compute second derivatives, we have:

- $f'' = -\exp(-x)$;
- $f'' = 20x^3$;
- $f'' = -\frac{2}{9}x^{-4/3}$;
- $f'' = 0$.

So the first and third are concave. The second is convex. The last is both since it is a straight line.

Solution to Q7.3.

We have to compute the expectation of

$$a(c+dZ) - b(c+dZ)^2.$$

The expectation of odd powers of Z is zero, of Z^2 is 1 and of Z^4 is 3. We therefore get

$$ac - b(c^2 + d^2).$$

Solution to Q7.4.

$$\log(2x) = \log 2 + \log x,$$

so the second and third are equivalent. The first and fifth differ by 1 so they are equivalent. The rest are not equivalent.

Solution to Q7.5

We have to compute the utility if we invest in the risk-free asset. This is

$$U(W_{rf}) = \log(1,005,000) = 13.8205,$$

in the up-state and the down-state:

$$U(W_{+}) = 13.8255, \quad U(W_{-}) = 13.8155.$$

The p for which we are indifferent gives the same expected utility and is equal to

$$\frac{U(W_{rf}) - U(W_{-})}{U(W_{+}) - U(W_{-})} = 0.50124.$$

So if p is greater than this, A would be bought.

Solution to Q7.6. We have to compute the utility if we invest in the risk-free asset. This is

$$U(W_{rf}) = 0.6340,$$

in the up-state and the down-state:

$$U(W_{+}) = 0.6358, \quad U(W_{-}) = 0.6321.$$

The p for which we are indifferent gives the same expected utility and is equal to

$$\frac{U(W_{rf}) - U(W_{-})}{U(W_{+}) - U(W_{-})} = 0.50125.$$

So if p is greater than this, A would be bought.

Solution to Q7.7. The investor is risk-neutral so he will invest in A if and only if the expectation is higher. This means $p > 0.5$.

Solution to Q7.8. We first compute the terminal wealth for each world-state and investment and get

X	Y	Z
1,040,000.00	1,050,000.00	1,070,000.00
1,070,000.00	1,050,000.00	1,030,000.00
1,000,000.00	1,010,000.00	1,010,000.00
1,080,000.00	1,080,000.00	1,080,000.00

.

We then compute the utility in each of these cases to get

X	Y	Z
13.854731	13.864301	13.883169
13.883169	13.864301	13.845069
13.815511	13.825461	13.825461
13.892472	13.892472	13.892472

.

The expected utilities are then

$$13.859044, \quad 13.85828295, \quad 13.85628724.$$

So we choose X.

Now for the square-root utility function. We compute the utility in each of the cases to get

X	Y	Z
1019.803903	1024.695077	1034.408043
1034.408043	1024.695077	1014.889157
1000.000000	1004.987562	1004.987562
1039.230485	1039.230485	1039.230485

.

The expected utilities are then

$$1022.12929, \quad 1021.689904, \quad 1020.690721$$

So we choose X.

Solution to Q7.9. When $X = 10$, the wealth the up-state is 10010 and in the down-state 9980. We compute

$U(W_u)$	9.211339872
$U(W_d)$	9.208338369
$\mathbb{E}(U(W_X))$	9.210589497

We exponentiate to get the indifference wealth of 10002.49156. The expected value is 2.5.

So the indifference price and the risk premium are, respectively,

$$0.008443838 \quad \text{and} \quad 2.491556162.$$

When $X = 100$, the wealth the up-state is 10100 and in the down-state 9900. We compute

$$\begin{array}{rl} U(W_u) & 9.220290703 \\ U(W_d) & 9.190137665 \\ \mathbb{E}(U(W_X)) & 9.212752443 \end{array}$$

We exponentiate to get the indifference wealth of 10024.14983. The expected value is 25.

So the indifference price and the risk premium are, respectively,

$$0.850173051 \quad \text{and} \quad 24.14982695.$$

When $X = 1000$, the wealth the up-state is $11,000$ and in the down-state $8,000$. We compute

$$\begin{array}{rl} U(W_u) & 9.305650552 \\ U(W_d) & 8.987196821 \\ \mathbb{E}(U(W_X)) & 9.226037119 \end{array}$$

We exponentiate to get the indifference wealth of 10158.20588. The expected value is 250.

So the indifference price and the risk premium are, respectively,

$$91.79411923 \quad \text{and} \quad 158.2058808.$$

Solution to Q7.11, 7.12, 7.13.

We can either use Taylor's theorem or write

$$V(W) = A + BW + CW^2,$$

and solve for the values of A, B and C that match $U(W_0), U'(W_0)$ and $U''(W_0)$.

We see that

$$C = 0.5U''(W_0),$$
$$B = U'(W_0) - 2CW_0,$$
$$A = U(W_0) - BW_0 - CW_0^2.$$

W_0	U	U'	U''	C	B	A
1	0	1	–1	–0.5	2	–1.5
10	2.302585093	0.1	–0.01	–0.005	0.2	0.802585093
100	4.605170186	0.01	–0.0001	–0.00005	0.02	3.105170186

Solution to Q7.13, 7.14, 7.15.

We compute

W_u	10010
W_d	9980
$U(W_u)$	–0.367511746
$U(W_d)$	–0.368615936
$\mathbb{E}(U(W_X))$	–0.367787793

The indifference wealth is computed by taking the log of the negative of the expected utility and dividing by $-a$. We get 10002.49156. The expected value is 2.5.

The indifference price is 2.491558282 and the risk premium is 0.008441718.

For the rest of 2.1, 2.2 and 2.3, we proceed similarly.

a	X	price	premium
1/10,000	10	2.491558282	0.008441718
1/10,000	100	24.1520394	0.847960602
1/10,000	1000	161.5097457	88.49025425
1/1,000	10	2.41520394	0.08479606
1/1,000	100	16.15097457	8.849025425
1/1,000	1000	–752.9119531	1002.911953
1/100,000	10	2.499156208	0.000843792
1/100,000	100	24.91558282	0.08441718
1/100,000	1000	241.520394	8.479606018

Solution to Q7.16. In each case, we compute the utility in the up and down states and compute the expected utility. We then work out the indifference price by inverting the utility function, and then computing the risk-premium.

For $X = 10$, the up and down wealths are 100020 and 99970.

up	–0.3678	11.5131	316.2594	46.4190	17.7837
down	–0.3680	11.5126	316.1803	46.4112	17.7815
Expected U	–0.3678	11.5130	316.2436	46.4174	17.7832
indiff. price	9.9980	9.9980	9.9990	9.9987	9.9985

For $X = 100$, the up and down wealths are 100200 and 99700.

up	–0.367	11.515	316.544	46.447	17.792
down	–0.369	11.510	315.753	46.369	17.769
Expected U	–0.368	11.514	316.386	46.431	17.787
indiff. price	99.800	99.800	99.900	99.867	99.850

For $X = 1000$, the up and down wealths are 102000 and 97000.

up	–0.3606	11.5327	319.3744	46.7233	17.8710
down	–0.3791	11.4825	311.4482	45.9470	17.6479
Expected U	–0.3643	11.5227	317.7892	46.5680	17.8264
indiff. Price	979.7999	979.7949	989.9482	986.5750	984.8842

Solution to Q7.17.

The first derivative must be positive and the second derivative negative. So we must have

$$ab > 0, \quad \text{and} \quad ab^2 < 0.$$

This means $b < 0$ and so $a < 0$.

Solution to Q7.18. Quadratic utility implies a mean–variance investor. We are being asked to describe the efficient frontier for an investor who borrows but does not place money on deposit.

We therefore get a half-line in weight space which starts at the tangent portfolio, v, and is of the form $-X$ units of the riskless asset and $1+X$ units of v. We also get a straight line from the minimal variance portfolio to the tangent portfolio with no risk-less holding.

In expected return/standard deviation space, we get an investment line tangent to the hyperbola at the tangent portfolio which goes through the riskless asset when extrapolated. So we have the part of the hyperbola from the riskless asset to the tangent portfolio and then a straight line going off to infinity.

Solution to Q7.19. We first have to compute the approximating quadratic utility function and then use this to compute the indifference price. To get the approximate utility, we match up to the second derivative at the current wealth. The approximation is

$$A + BW + CW^2.$$

We find

$$C = -1.8394E-09,$$
$$B = 7.35759E-05,$$
$$A = -0.919698603.$$

When $X = 10$, we find

utility up	–0.367144418
utility down	–0.368247505
expected utility	–0.367971733

We then solve a quadratic to match the wealth. This has two solutions. We take the one which is close to the initial wealth and get 9997.491565.

We therefore have

indiff. price	–2.508435388
risk-neutral price	–2.5
risk premium	0.008435388

With $X = 100$, we proceed similarly

utility up	–0.360595428
utility down	-0.37157663
expected utility	–0.368831329
indiff. W	9974.158389
indiff price	–25.84161056
risk-neutral price	–25
risk premium	0.841610558

With $X = 1,000$, we proceed similarly

utility up	–0.301661142
utility down	–0.406506782
expected utility	–0.380295372
indiff, W	9668.01084
indiff price	–331.9891599
risk-neutral price	–250
risk premium	81.98915989

Note that although the answer is given here in full detail, it could be found more simply by working with the equivalent quadratic utility function

$$W + (C/B)W^2.$$

Chapter 8

Solution to Q8.1. First, we compute for $U(W) = -W^{\alpha}$. We have

$$U'(W) = -\alpha W^{\alpha-1},$$
$$U''(W) = -\alpha(\alpha-1)W^{\alpha-2},$$

so

$$A(W) = -\frac{U''(W)}{U'(W)} = \frac{(1-\alpha)}{W},$$

and

$$R(W) = WA(W) = 1-\alpha.$$

We now set $\alpha = -1/4$, to get $(5/4)/W$ and $5/4$.
We set $\alpha = -1/2$, to get $(3/2)/W$ and $3/2$.
We set $\alpha = -1/3$, to get $(4/3)/W$ and $4/3$.
For $U(W) = 1 - e^{-W}$, we find

$$U'(W) = e^{-W},$$
$$U''(W) = -e^{-W},$$

so

$$A(W) = 1, \quad R(W) = W.$$

Solution to Q8.2. If $U(W) = -1/W$, we have

$$U'(W) = W^{-2}, \quad U''(W) = -2W^{-3},$$

so

$$A(W) = 2/W, \; R(W) = 2.$$

This utility has decreasing absolute risk aversion but constant relative risk aversion. It is an increasing concave function.

If $U(W) = W - aW^2$, we have

$$U'(W) = 1 - 2aW, \quad U''(W) = -2a.$$

So

$$A(W) = \frac{2a}{1-2aW}, \; R(W) = \frac{2aW}{1-2aW}.$$

Provided $W < 1/(2a)$, we have an increasing concave utility function. The absolute risk aversion is increasing with W and so does the relative risk aversion. This is not very realistic. We also have decreasing utility past $1/(2a)$ which is not at all realistic.

If $U(x) = x$, both risk aversions are zero. This gives a risk-neutral investor.

Solution to Q8.3. The absolute risk aversion is defined by

$$A(W) = -\frac{U''(W)}{U'(W)}.$$

For these three, we have

$$A(W) = \frac{3}{4W},$$
$$A(W) = \frac{1}{W},$$
$$A(W) = 1.$$

The absolute risk aversion should decrease with wealth since a rich person is a lot less concerned about losing small sums of money.

Solution to Q8.4. We can use the given behaviours to compute bounds on the absolute risk aversion. We know the indifference price is

$$\pi = \mathbb{E}(X) - \frac{A}{2}\sigma_X^2.$$

If the investor refuses then the indifference price is negative. So in the first case we have

$$100 - A\frac{100^2}{2} < 0,$$

so $A > 0.02$.

In the second case,

$$100 - A\frac{50^2}{2} > 0,$$

so

$$A < 0.08.$$

We can now compute π using both these values for the five investments to get bounds on the indifference prices. If positive we would do the investment, if negative we would not.

We compute

$\mathbb{E}(X)$	σ	var	$\pi_{\min}$	$\pi_{\max}$
200	50	2500	175	100
200	100	10000	100	–200
200	150	22500	–25	–700
200	200	40000	–200	–1400
200	250	62500	–425	–2300

So the first investment would probably be acceptable, for the second there is insufficient data, and for the others we would not.

Solution to Q8.5. We first compute bounds on $A(W_0)$ with $W_0 = 1,000,000$. If the investment is accepted we have

$$\pi > 0 \implies A(W_0) > \frac{2\mathbb{E}(X)}{\sigma_X^2}.$$

So we have $A(W_0) < \frac{1}{100}$.

Similarly, we have $A(W_0) > \frac{1}{400}$.

These imply

$$\frac{1000000}{400} < R(W) < \frac{1000000}{100}.$$

We have that $R(W)$ is constant so these will be the same for $W_1 = 100000$. We conclude

$$\frac{1}{40} < A(W_1) < \frac{1}{10}.$$

The indifference price $\pi(X)$ will therefore satisfy

$$25 - \frac{1}{2}\frac{X^2}{10} < \pi(X) < 25 - \frac{1}{2}\frac{X^2}{40}.$$

So if

$$25 - \frac{1}{2}\frac{X^2}{40} < 0$$

then the investment would not be done. That is if

$$X > \sqrt{80 \times 25}.$$

It would be done if

$$25 - \frac{X^2}{20} > 0,$$

i.e. if

$$X < \sqrt{500}.$$

In between, we would not be able to say.

Solution to Q8.6. The first investor is mean–variance. The means are

$$4, \quad 3.33, \quad 3.67.$$

The variances are

$$98.67, \quad 172.22, \quad 131.05.$$

Strategy A has highest mean and lowest variance so best, followed by mix, followed by B.

The expected log utilities are 0.034426, 0.024184, 0.029537 so A then a mix then B.

The third utility function is equivalent to the second one so the same answer.

Solution to Q8.7. An investor with exponential utility has a constant absolute risk aversion so behaviour will not change on a fall in wealth.

So for $\sigma \leq 100$ will accept and $\sigma \geq 200$ will reject.

For the power utility investor, there is constant relative risk aversion so absolute risk-aversion will multiply by 10. It is the square of the standard deviation that is important so we must divide the change points by $\sqrt{10}$. So he accepts for σ less than

$$100/\sqrt{10}$$

and rejects for σ greater than $200/\sqrt{10}$.

Chapter 9

Solution to Q9.1. The first bet is the example bet with double the stake. He will therefore not do this.

The second bet is worse than the example so no again.

Solution to Q9.2. A is rational since he is a risk-neutral investor. We take $U(W) = W$.

B is not since a cash-sum will always be preferred to any investment with a positive standard deviation. This contradicts "certainty equivalence." For example, B would prefer -1 dollars to an investment that pays $1,000,000$ dollars 99% of the time and zero otherwise.

Chapter 10

Solution to Q10.1.

For P, we compute $\log(1+r)$ in the three states to get

$$0.009950331, \quad 0.039220713, \quad 0.086177696.$$

The probability weighted sum gives us the expectation which is

$$\mathbb{E}(1+R_P) = 0.043642363.$$

For Q, we compute $\log(1+r)$ in the three states to get

$$0.019802627, \quad 0.039220713, \quad 0.067658648.$$

The probability weighted sum gives us the expectation which is

$$\mathbb{E}(1+R_Q) = 0.043868477.$$

So we conclude that Q is preferred on geometric means.

Solution to Q10.2. For P, we compute $\log(1+r)$ in the three states to get

$$0.019802627, \quad 0.048790164, \quad 0.058268908.$$

The probability weighted sum gives us the expectation which is

$$\mathbb{E}(1+R_P) = 0.043912966.$$

For Q, we compute $\log(1+r)$ in the three states to get

$$0.019802627, \quad 0.039220713, \quad 0.067658648.$$

The probability weighted sum gives us the expectation which is

$$\mathbb{E}(1+R_Q) = 0.043868477.$$

So we conclude that P is preferred on geometric means.

Solution to Q10.3. For P, we compute $\log(1+r)$ in the three states to get

$$0.019802627, \quad 0.048790164, \quad 0.067658648.$$

The probability weighted sum gives us the expectation which is

$$\mathbb{E}(1+R_P) = 0.055325653.$$

For Q, we compute $\log(1+r)$ in the three states to get

$$0.019802627, \quad 0.039220713, \quad 0.067658648.$$

The probability weighted sum gives us the expectation which is

$$\mathbb{E}(1+R_Q) = 0.043868477.$$

So we conclude that P is preferred on geometric means.

Solution to Q10.4. We compute

	R_X	$\log(1+R_X)$
0.2	6	0.058268908
0.4	8	0.076961041
0.3	9	0.086177696
0.1	11	0.104360015

	R_Y	$\log(1+R_Y)$
0.1	7	0.067658648
0.3	8	0.076961041
0.2	9	0.086177696
0.3	10	0.09531018
0.1	11	0.104360015

	R_Z	$\log(1+R_Z)$
0.4	8	0.076961041
0.3	9	0.086177696
0.2	10	0.09531018
0.1	11	0.104360015

We take the probability weighted sums to get $\mathbb{E}(\log(1+r))$ equal to

$$0.078727508, \quad 0.086118772 \text{ and } 0.086135763.$$

This says that Z will be preferred for very long term investors.

For fixed-time-horizon investors, we look at $\mathbb{E}(r)$ and get

$$8.2, \quad 9, \quad \text{and} \quad 9.$$

So either Y or Z would be chosen.

Solution to Q10.5. If r has positive chance of losing all money then it has zero geometric mean. So s will have higher geometric mean.

Solution to Q10.6. Since the expected return of the stock is greater than that of borrowing, to maximise for a specific date we go long as many stocks as possible and borrow to fund it.

For very long term investment, we have to find the value of X that maximises

$$\mathbb{E}(\log(1+r_s X + r_B(1-X))).$$

That is,

$$0.6\log(1+0.08X+0.02)+0.4\log(1+0.02-0.07X).$$

We have to find the place where the derivative of this is zero. Differentiating, we get

$$\frac{0.6\times 0.08}{1.02+0.08x} - \frac{0.4\times 0.07}{1.02-0.07x} = 0.$$

Multiplying through by the product of the denominators and collecting terms we get

$$0.0204 - 0.0056x = 0.$$

So

$$x = 3.6428571.$$

Sketching the graph it is clear that this is the unique maximum (or compute second derivatives.)

Solution to Q10.7. Since the expected return of the stock is greater than that of borrowing, to maximise for a specific date we go long as many stocks as possible and borrow to fund it.

For very long term investment, we have to find the value of X that maximises the $\mathbb{E}(\log(1+r))$. We have to consider what happens for $X<1$ and $X>1$ separately since different borrowing and lending rates.

First if $X=0$, we get

$$\log(1.02) = 0.0198026.$$

If $X=1$ we get

$$0.5\log(1.25)+0.5\log(0.9) = 0.058891518.$$

We have to differentiate

$$0.5\log(1.02+0.23X)+0.5\log(1.02-0.12X)$$

to find whether there is a maximum on $X<1$. A maximum would occur if

$$\frac{0.23}{1.02+0.23X} - \frac{0.12}{1.02-0.12X} = 0.$$

Solving for X shows that the derivative is never zero for $X<1$ and so the maximum occurs at $X=1$.

On $X > 1$, we have to find the maximum again, we now have to solve

$$\frac{0.2}{1.05 + 0.2X} - \frac{0.15}{1.05 - 0.15X} = 0.$$

The solution is 0.875 and so there is no maximum on $X \geq 1$.

An investor for the very long term would therefore hold the stock and zero cash.

Chapter 11

Solution to Q11.1.

We can compute tables and we get

r	f_x	F_X	$\int F_X$
0	0	0	0
1	0.25	0.25	0
2	0	0.25	0.25
3	0	0.25	0.5
4	0.5	0.75	0.75
5	0	0.75	1.5
6	0	0.75	2.25
7	0	0.75	3
8	0	0.75	3.75
9	0.25	1	4.5
10	0	1	5.5

r	f_Y	F_Y	$\int F_Y$
0	0	0	0
1	0	0	0
2	0.2	0.2	0
3	0	0.2	0.2
4	0.5	0.7	0.4
5	0	0.7	1.1
6	0	0.7	1.8
7	0.3	1	2.5
8	0	1	3.5
9	0	1	4.5
10	0	1	5.5

The differences are

$F_X - F_Y$	$\int F_X - \int F_Y$
0	0
0.25	0
0.05	0.25
0.05	0.3
0.05	0.35
0.05	0.4
0.05	0.45
−0.25	0.5
−0.25	0.25
0	0
0	0

So Y is second-order stochastically dominant to X but not first-order stochastically dominant.

Solution to Q11.2. We can compute tables and we get

r	f_x	F_X	$\int F_X$
0	0	0	0
1	0	0	0
2	0.25	0.25	0
3	0	0.25	0.25
4	0	0.25	0.5
5	0.5	0.75	0.75
6	0.25	1	1.5
7	0	1	2.5
8	0	1	3.5
9	0	1	4.5
10	0	1	5.5

r	f_Y	F_Y	$\int F_Y$
0	0	0	0
1	0	0	0
2	0.2	0.2	0
3	0	0.2	0.2
4	0.5	0.7	0.4
5	0	0.7	1.1
6	0	0.7	1.8
7	0.3	1	2.5
8	0	1	3.5
9	0	1	4.5
10	0	1	5.5

The differences are

$F_X - F_Y$	$\int F_X - \int F_Y$
0	0
0	0
0.05	0
0.05	0.05
–0.45	0.1
0.05	–0.35
0.3	–0.3
0	0
0	0
0	0
0	0

So in this case, we can say nothing.

Solution to Q11.3. We can compute tables to get

r	f_x	F_X	$\int F_X$
0	0	0	0
1	0	0	0
2	0.1	0.1	0
3	0	0.1	0.1
4	0	0.1	0.2
5	0.4	0.5	0.3
6	0	0.5	0.8
7	0.5	1	1.3
8	0	1	2.3
9	0	1	3.3
10	0	1	4.3

r	f_Y	F_Y	$\int F_Y$
0	0	0	0
1	0	0	0
2	0.2	0.2	0
3	0	0.2	0.2
4	0.5	0.7	0.4
5	0	0.7	1.1
6	0	0.7	1.8
7	0.3	1	2.5
8	0	1	3.5
9	0	1	4.5
10	0	1	5.5

The differences are

$F_X - F_Y$	$\int F_X - \int F_Y$
0	0
0	0
–0.1	0
–0.1	–0.1
–0.6	–0.2
–0.2	–0.8
–0.2	–1
0	–1.2
0	–1.2
0	–1.2
0	–1.2

We have first-order stochastic dominance of X over Y. Note that since first order implies second order, it would be a waste of time to compute the table for second order.

Solution to Q11.4. We create tables of cumulative and integral of cumulatives for each investment.

r	f_x	F_X	$\int F_X$
5	0	0	0
6	0.2	0.2	0
7	0	0.2	0.2
8	0.4	0.6	0.4
9	0.3	0.9	1
10	0	0.9	1.9
11	0.1	1	2.8
12	0	1	3.8
13	0	1	4.8

r	f_Y	F_Y	$\int F_Y$
5	0	0	0
6	0	0	0
7	0.1	0.1	0
8	0.3	0.4	0.1
9	0.2	0.6	0.5
10	0.3	0.9	1.1
11	0.1	1	2
12	0	1	3
13	0	1	4

r	f_Z	F_Z	$\int F_Z$
5	0	0	0
6	0	0	0
7	0	0	0
8	0.4	0.4	0
9	0.3	0.7	0.4
10	0.2	0.9	1.1
11	0.1	1	2
12	0	1	3
13	0	1	4

We also compute the expectations; we get 8.2, 9, 9. This means that Y and Z may be first-order stochastically dominant to X but will not be to each other.

We therefore compute the difference of cumulatives to test first-order stochastic dominance for the two possible pairs.

	$F_x - F_y$	$F_x - F_z$
5	0	0
6	0.2	0.2
7	0.1	0.2
8	0.2	0.2
9	0.3	0.2
10	0	0
11	0	0
12	0	0
13	0	0
min	0	0
max	0.3	0.2

We see that Y and Z are both first-order stochastically dominant to X and it follows that they are also second-order stochastically dominant. So Y and Z will be preferred to X by an investor with a strictly increasing utility function.

We test Y and Z for second-order stochastic dominance.

	$\int F_y - \int F_z$
5	0
6	0
7	0
8	0.1
9	0.1
10	0
11	0
12	0
13	0
min	0
max	0.1

So Z has second-order stochastic dominance over Y. So an investor with an increasing concave utility function will prefer Z then Y then X.

Solution to Q11.5. For risk-neutrality, we simply need higher expectation. The means are

$$0.5(a_j + b_j).$$

So A_1 is preferred to A_2 if and only if

$$a_1 + b_1 > a_2 + b_2.$$

For increasing utility functions, we must test first-order stochastic dominance. The cumulative distribution functions are

$$\frac{1}{b_j - a_j}\left((x - a_j)_+ - (x - b_j)_+\right).$$

Note that y_+ means y when $y > 0$ and 0 otherwise. That is, they are zero below a_j and increasing with slope $1/(b_j - a_j)$ until reach 1 at b_j and are then constant.

If A_1 is first-order stochastically dominant over A_2, then we must have $a_1 \geq a_2$ or its cumulative will be greater than that of A_2 at $(a_1 + a_2)/2$. Similarly, we must have $b_1 \geq b_2$ or between b_1 and b_2 its cumulative will be 1 whilst that of A_2 is less than 1.

If $a_1 = a_2$ and $b_1 = b_2$ then the two functions are the same. So for first-order stochastically dominant we must have $a_1 \geq a_2$, $b_1 \geq b_2$ and at least one of $a_1 > a_2$ and $b_1 > b_2$ must hold. But if these hold then first-order stochastic dominance does too.

Solution to Q11.6. For risk-neutral investors, we must have a higher mean so we need

$$a_1 + b_1 > a_2 + b_2.$$

For increasing utility, the question is first-order dominance. The cumulative distribution function is zero below a_j then 0.5 up to b_j and then 1.

Clearly, we must have $a_1 \geq a_2$ or A_2 will have lower cumulative just above a_1. Similarly, we must have $b_1 \geq b_2$ or just above b_1, A_2 will have lower cumulative. If $a_1 = a_2$ and $b_1 = b_2$ then the distributions are the same. So we take $a_1 > a_2$ or $b_1 > b_2$.

We then have that the cumulative for A_2 minus that of A_1 is: zero for $x < a_2$; 0.5 for $x \in [a_2, a_1]$; zero on $[a_1, b_2]$; 0.5 on $[b_2, b_1]$; and zero above b_1. So first-order stochastic dominance does hold.

For increasing concave utility, we need to check second-order stochastic dominance. Either first-order stochastic dominance holds, and we are done, or it does not and we need to analyse.

So suppose A_1 is not first-order stochastically dominant to A_2. Suppose also that the investments are not identical. We must have

$$a_1 + b_1 \geq a_2 + b_2$$

by our general result on means under second-order stochastic dominance.

If not first-order stochastically dominant, either $a_1 < a_2$ or $b_1 < b_2$. If $a_1 < a_2$, the integral of the cumulative of A_1 is positive and so greater than that of A_2 on $[a_1, a_2]$ hence not second-order stochastically dominant.

So if second-order stochastically dominant without being first order, we must have $a_1 \geq a_2$ and $b_1 < b_2$. So $a_2 \leq a_1 < b_1 < b_2$.

The difference of the integral of the cumulatives is one half of

$$(x - a_2)_+ - (x - a_1)_+ - (x - b_1)_+ + (x - b_2)_+.$$

Note that y_+ means y when $y > 0$ and 0 otherwise. Since the function is a straight line except at the points a_2, a_1, b_1, b_2, we need to check its values at these four points.

- At a_2, it is zero.
- At a_1, it is $a_1 - a_2$.
- At b_1 it is $b_1 - a_2 - b_1 + a_1 = a_1 - a_2$.
- At b_2 and above, it is $-a_2 + a_1 + b_1 - b_2$.

So we must have

$$a_2 \le a_1$$

and

$$a_1 + b_1 \ge a_2 + b_2.$$

One of these inequalities must be strict.

Solution to Q11.7. We have to translate the behaviours into mathematical statements for each of the cases.

For very long term and log utility, we always get the same answers: we have to compute $\mathbb{E}(\log(1+r))$.

We have the table of logs:

$\log(1+r_X)$	$\log(1+r_Y)$	$\log(1+r_Z)$
0.048790164	0.058268908	0.048790164
0.067658648	0.039220713	0.048790164
–0.010050336	–0.020202707	0.048790164
0.104360015	0.173953307	0.048790164

We get the expectations:

$$0.048239595, \quad 0.048428566, \quad 0.048790164$$

so Z followed by Y followed by X.

The increasing linear utility means a risk-neutral investor so we do expectations and get

$$5, \quad 5.1, \quad 5$$

so Y is best followed by X and Z equally.

For mean–variance investors we need the variances which are

$$12, \quad 30.09, \quad 0.$$

So Z is better than X since same mean and lower variance. We can say nothing about Y since it has higher mean and variance.

For the last two, we need to compare using first- and second-order stochastic dominance.

If first-order stochastically dominant, we must have higher mean. The only possibilities are therefore that Y is first-order stochastically dominant to one of X and Z. However, its cumulative is positive at -1.5 where theirs are not, so no first-order stochastic dominance.

For second-order stochastic dominance, the same argument shows that Y

cannot be stochastically dominant to X or Z. So the only question regards X and Z. Since they have the same mean, and Z has zero variance this is reasonably likely.

The integral of the cumulative of Z is the function

$$(x-5)_+$$

where $f_+(x)$ is zero if $f(x) < 0$ and equal to $f(x)$ otherwise.

The integral of the cumulative of X is

$$0.2(x+1)_+ + 0.4(x-5)_+ + 0.3(x-7)_+ + 0.1(x-11)_+.$$

For $x \leq -1$, both are zero. For $x \geq 11$, both equal

$$x-5.$$

At $x=5$, the integral of cumulative of X is bigger than zero so bigger than that of Z. At $x=7$ they are equal to 2.4 and 2 so X is bigger again.

Since the functions are straight lines except at these four points, we conclude that Z is second-order stochastically dominant to X.

Chapter 12

Solution to Q12.1. The probability of **not** getting an excess in N days is

$$0.99^N.$$

So we need to find the smallest integer N such that

$$0.99^N < 0.5.$$

Taking logs and dividing, we need

$$N > 68.9$$

so $N = 69$.

For the second one, we need

$$0.99^N < 1 - 0.99 = 0.01.$$

So this requires

$$N > 458.2$$

so $N = 459$.

For the last part, take

$$1 - 0.99^{10}.$$

Solution to Q12.2. The probability of zero excesses is

$$(1-0.01)^{10} = 0.90438.$$

Solution to Q12.3. The probability of zero excesses is

$$(1-0.05)^{90} = 0.00989.$$

The probability of one excess is

$$90 \times (1-0.05)^{89} \times 0.05 = 0.04684.$$

The sum of these is

$$0.05673.$$

Solution to Q12.4 and 12.5. For zero excesses, we have to compute

$$0.95^{50}, \qquad 0.99^{100}.$$

These equal

$$0.076944975, \qquad 0.005920529.$$

For more than 2, we compute for 0, 1, 2, and then subtract from 1.
For 1 we get

$$50 \times 0.95^{49} \times 0.05 = 0.20249 \quad \text{and} \quad 0.03116.$$

For 2, we get

$$50 \times 49/2 \times 0.95^{48} \times 0.05^2 = 0.26110 \quad \text{and} \quad 0.081182.$$

Summing, we get

$$0.5405, \qquad 0.1183.$$

Subtracting from 1, we get

$$0.4595, \qquad 0.8817.$$

Solution to Q12.6. For the 1% level, we compute

$$N^{-1}(0.01) = -2.326.$$

The value at the 1% is then

$$110 + 10 \times (-2.326) = 86.737.$$

So the VAR at that level is 13.263.

For the 5% level, we compute

$$N^{-1}(0.05) = -1.645.$$

The value at the 5% level is then

$$110 + 10 \times (-1.645) = 93.551.$$

So the VAR at that level is 6.449.

Solution to Q12.7. We compute probabilities to get

$n = 0$	0.404731973
$n = 1$	0.367938157
$n = 2$	0.165386343
$n = 3$	0.049003361
$n <= 2$	0.938056473
$n <= 3$	0.987059834
$n > 2$	0.061943527
$n > 3$	0.012940166

More than two excesses would be a cause for some concern but it is perfectly possible since 6% probability events are not that rare. More than 3 is more worrying since we are down to 1% levels. We can expect 2 quite often so need not worry at all.

Solution to Q12.8. The 0.1% VAR is 150,000,000.

For VAR at $p > 0.11\%$ the value at that p is

$$900 + 300\frac{p-q}{1-q}$$

where $q = 0.0011$.

We plug in p and subtract from 1000 to get 97.3 and 85.3.

The change would not affect the VAR.

Chapter 13

Solution to Q13.1. We compute using the CAPM formula to get 0.21.

Solution to Q13.2. Since there is no diversifiable risk, we must hold a multiple of the market portfolio and we compute that the beta is 0.75. We compute using the CAPM formula to get an expected return of 0.15.

Solution to Q13.3. We compute to get 15, 3.

Solution to Q13.4. We compute using the expression for the return of a portfolio in terms of covariance with the tangent portfolio to get 7.6.

Solution to Q13.5. We compute using the expression for the return of a portfolio in terms of covariance with the tangent portfolio to get that the expected return of X is 14.

We therefore hold $5/3$ units of X and $-2/3$ units of the risk-free asset.

Solution to Q13.6. In two-factor CAPM, the market portfolio is always efficient and the zero-beta portfolio never is.

We compute that the gain for each unit of beta is

$$\frac{20-10}{1.5-0.5} = 10,$$

and so

$$s = 10 - 0.5 \times 10 = 5.$$

So the market's return is $5 + 10 = 15$ and the zero beta portfolio has return 5.

Solution to Q13.7. The zero-beta return is 5 and the gain per unit of beta is 10. So we get 10 and 25.

Solution to Q13.8. We compute to get 1.5 and -0.5.

Solution to Q13.9. The two-factor CAPM ensures that if two securities have the same beta then they also have the same α. That is α can be written as a function of beta.

Solution to Q13.10. The two assumptions are independent. Neither can hold or both or either one.

Solution to Q13.11. The betas are 1 and 0.5. So the gain per unit of beta is 5 and the zero beta return is 5. So the return of the market is therefore 10. C has a beta of 0.48 and so has return 7.4.

Solution to Q13.12. The return is $4+6\beta$. So β is $(20-4)/6=8/3$ and $-4/6=-2/3$.

Chapter 14

Solution to Q14.1. With zero units of Z risk, we get a constant of 3, and with one unit we get 5, so we get 2 for each unit of Z risk. Hence we have

$$\begin{aligned}\alpha_1 &= 3+2\times 2=7,\\ \alpha_2 &= 3-2=1,\\ \alpha_3 &= 3+1.5\times 2=6.\end{aligned}$$

Solution to Q14.2. The portfolio consisting of 2 units of A and -1 units of B has the constant return -1 so this is the risk-free rate.

Solution to Q14.3.

- This is a true instant arbitrage. We simply buy the shares, exercise the options and sell the beans immediately.
- This is a dynamic arbitrage. It is a true arbitrage insofar as the model is correct. (It's called Black–Scholes–Merton!)
- This is a statistical arbitrage; it may work but it is not true arbitrage.

Solution to Q14.4. We solve the matrix problem

$$\begin{pmatrix}1 & 2 & 1\\ 1 & 3 & 1\\ 1 & 1 & 2\end{pmatrix}\begin{pmatrix}r_f\\ \mu_1\\ \mu_2\end{pmatrix}=\begin{pmatrix}13\\ 14\\ 20\end{pmatrix}$$

and we get

$$R_f=3,\mu_1=1,\mu_2=8.$$

Solution to Q14.5. We solve the matrix problem

$$\begin{pmatrix} 1 & 2 & 1 \\ 1 & 1 & 2 \\ 1 & 0 & 1 \end{pmatrix} \begin{pmatrix} r_f \\ \mu_1 \\ \mu_2 \end{pmatrix} = \begin{pmatrix} 20 \\ 14 \\ 3 \end{pmatrix}$$

and we get

$$R_f = 0.5, \mu_1 = 8.5, \mu_2 = 2.5.$$

Solution to Q14.6. We solve the matrix problem

$$\begin{pmatrix} 1 & 3 & 1 \\ 1 & 1 & 2 \\ 1 & 2 & -1 \end{pmatrix} \begin{pmatrix} r_f \\ \mu_1 \\ \mu_2 \end{pmatrix} = \begin{pmatrix} 28 \\ 22 \\ 16 \end{pmatrix}$$

and we get

$$R_f = 10, \mu_1 = 4.8, \mu_2 = 3.6.$$

Solution to Q14.7. We solve the matrix equation

$$\begin{pmatrix} 0.5 & 0.5 \\ 1.2 & 0.2 \end{pmatrix} \begin{pmatrix} \mu_1 \\ \mu_2 \end{pmatrix} = \begin{pmatrix} 10-5 \\ 12-5 \end{pmatrix}.$$

This has the solution

$$\begin{pmatrix} 5 \\ 5 \end{pmatrix}.$$

The expected returns are then

$$5 + 0.8 \times 5 + 0.2 \times 5 = 10,$$
$$5 + 0.2 \times 5 + 0.8 \times 5 = 10.$$

Solution to Q14.8. We solve the matrix equation

$$\begin{pmatrix} 0.75 & 0.25 \\ 2 & 2 \end{pmatrix} \begin{pmatrix} \mu_1 \\ \mu_2 \end{pmatrix} = \begin{pmatrix} 10-4 \\ 18-4 \end{pmatrix}.$$

This has the solution

$$\begin{pmatrix} 8.5 \\ -1.5 \end{pmatrix}.$$

The expected returns are then

$$4+1\times 8.5-1\times(-1.5)=14,$$
$$4+0.5\times 8.5+0.5\times(-1.5)=7.5.$$

Chapter 15

Solution to Q15.1. Someone has to win the first year. Call this person X. The probability they win again is $1/N$. The probability of X winning k times is $(1/N)^{k-1}$ given that they won the first time. Somebody has to win the first time so this is the answer.

Solution to Q15.2. Strong efficiency does not mean that Polly will always adopt an equally good strategy for winning. Strong efficiency does not rule out the possibility that well-diversified portfolios offer a better risk/return trade-off than ones that are not. Second, our objective here is not to invest well but win a competition. The strategy to win is to maximise the probability of doing best without worrying about risk so all losing portfolios are equally poor. So to maximise winning chances the best bet would be to put all the money into the stock with the highest apparent risk.

Solution to Q15.3. Obviously, if such a strategy truly worked it would contradict weak efficiency. However, backtesting on the same set it was trained on shows nothing. So the first question is what data was used to develop the strategy and what was it tested on? Has it been tested on new data that arrived after it was developed? Does it use mid-prices or real buy and sell prices? Could unexpected events cause large losses and how likely are such events?

Solution to Q15.4. The first study suggests that using privately held information does help with trading so markets are not strongly efficient.

The second study suggests that using publicly held information does not help with trading and so is consistent with semi-strong efficiency.

Chapter 16

Solution to Q16.1. This is the probability that $\sqrt{2}Z > 2$ with Z normal and so equals $N(-\sqrt{2})$.

Solution to Q16.2. The two conditions are independent and the second is of probability 0.5 so it is simply

$$\frac{1}{2}N(-1).$$

Solution to Q16.3. For any such partition, the expectation will be $b-a$ for all n. This follows immediately from the definition. So it certainly converges to $b-a$.

If the elements of P_n are denoted p_j, the variance will be

$$2\sum_{i=0}^{k-1}(p_{j+1}-p_j)^2.$$

If $|P_n| \leq \delta$, then this is less than or equal to

$$2\delta\sum_{i=0}^{k-1}(p_{j+1}-p_j) = 2\delta(b-a).$$

This clearly goes to zero as $\delta \to 0$.

Solution to Q16.4. The variance at time t is

$$t^{2\alpha+\beta}.$$

So we must have

$$2\alpha+\beta = 1.$$

Looking at the increments from 1 we see that we must have

$$t-1 = \mathbb{E}((t^{\alpha}W_{t^{\beta}} - W_1)^2) = t^{2\alpha+\beta} - 2t^{\alpha}\min(t^{\beta},1)+1.$$

For $\beta \geq 1$, this yields

$$t-1 = t^{2\alpha+\beta} - 2t^{\alpha}+1.$$

and so

$$t-1 = t-2t^{\alpha}+1.$$

This must hold for all t so $\alpha = 0$ and $\beta = 1$.

If $\beta < 1$, then we have

$$t - 1 = t - 2t^{\alpha+\beta} + 1.$$

So $\alpha + \beta = 0$ and $\alpha = 1$ and $\beta = -1$.

So our only additional possibility is

$$X_t = tW_{1/t}$$

and one can check that this works.

Solution to Q16.5. For the first one, this is just the probability

$$\mathbb{P}\left(\sqrt{j}Z > j\right) = \mathbb{P}\left(Z > \sqrt{j}\right) = N\left(-\sqrt{j}\right),$$

with Z a standard normal.

For the second one, we write $W_1 = X$, $W_2 = X + Y$ with X, Y standard normals. The event is then $X > 0$ and $X + Y < 0$. The probability of $X > 0$ is 0.5. Conditional on its occurrence, we need $Y < 0$ and $|Y| > |X|$. Each of these has probability 0.5 and are independent, so the answer is $1/8$.

For the third, let W_t be a standard Brownian motion. As usual write

$$W_t = W_s + (W_t - W_s).$$

So

$$\begin{aligned}
\mathbb{E}(W_t^k \mid W_s) &= \mathbb{E}((W_s + (W_t - W_s))^k \mid W_s), \\
&= \sum_{j=0}^{k} \binom{k}{j} W_s^{k-j} \mathbb{E}((W_t - W_s)^j), \\
&= \sum_{j=0}^{k} \binom{k}{j} W_s^{k-j} (t-s)^{j/2} \mathbb{E}(Z^j),
\end{aligned}$$

with Z standard $N(0,1)$. We have

$$\mathbb{E}(Z^2) = 1,\ \mathbb{E}(Z^4) = 3.$$

The odd powers are zero. So

$$\begin{aligned}
\mathbb{E}(W_t^4 \mid W_s) &= W_s^4 + 6W_s^2(t-s) + 3(t-s)^2, \\
\mathbb{E}(W_t^3 \mid W_s) &= W_s^3 + 3W_s(t-s), \\
\mathbb{E}(W_t^2 \mid W_s) &= W_s^2 + t - s.
\end{aligned}$$

Solution to Q16.6. First, note that

$$X_t = W_T - W_{T-t}$$

starts at zero. We certainly have

$$\mathbb{E}(X_t) = 0$$

and

$$\text{Var}(X_t) = \mathbb{E}((W_{T-t} - W_T)^2) = T - t,$$

Since $W_T = W_{T-t} + Y_t$ with Y_t an independent normal of variance Y_t.

The continuity of the sample paths is clear.

We need to show that for any t and s with $t > s$,

$$X_t - X_s = W_{T-s} - W_{T-t}$$

has variance $t - s$, but from the definition of Brownian motion, it is $T - s - T + t = t - s$.

The independence of increments property remains. However, this is just a rephrasing of the fact that the behaviour up to time t is independent of that after time t.

Solution to Q16.7. We need to compute

$$\text{Cov}(at + \sigma W_t, as + \sigma W_s) = \text{Cov}(\sigma W_t, \sigma W_s) = \sigma^2 \min(s,t).$$

Solution to Q16.8. We flip the path about its value at time 1. The new path is a Brownian motion. The normality of increments follows from the fact that a negative of a normal is a normal with the same properties. The fact that the variance is $t - s$ can be proven when $s < 1 < t$ be decomposing into two pieces about from s to 1 and 1 to t. Continuity of sample paths and independence is clear.

Solution to Q16.9. If we divide into N equal pieces of size T/N then the quadratic variation for that partition is

$$\sum_{j=0}^{N-1} |f((j+1)T/N) - f(jT/N)|^2 \leq \sum_{j=0}^{N-1} C^2 |1/N|^{2\gamma}.$$

This is less than or equal to

$$C^2 N \frac{1}{N^{2\gamma}} = C^2 N^{1-2\gamma}.$$

This goes to zero as $N \to +\infty$. If we set $N = 2^k$, we see that the dyadic quadratic variation is zero.

Solution to Q16.10. We have

$$S_t = S_0 e^{(\mu - 0.5\sigma^2)t + \sigma W_t},$$

so

$$S_t^k = S_0^k e^{k(\mu - 0.5\sigma^2)t + \sigma k W_t},$$

and we therefore get a geometric Brownian motion.

Solution to Q16.11. We simply compute

$$\frac{\nu - r}{\sigma}$$

to get

$$0.5, \quad 0.35, \quad -0.025, \quad -0.1.$$

Solution to Q16.12. We are given ν so we first compute

$$\mu = \nu - 0.5\sigma^2 = 0.095.$$

We then just plug into the formulas for the means of log-normals to get

$$110.5170918, \quad 122.1402758, \quad 134.9858808$$

and the variances to get

$$122.753018, \quad 301.3685798, \quad 554.9177887.$$

Solution to Q16.13. We are given ν so we first compute

$$\mu = \nu - 0.5\sigma^2 = 0.055.$$

We then just plug into the formulas for being above the level to get

$$0.703514298, \quad 0.694136423, \quad 0.698575907.$$

Solution to Q16.14. We are given ν so we first compute

$$\mu = \nu - 0.5\sigma^2 = 0.03.$$

We then just plug into the formulas for being above the level to get

$$0.372003793, \quad 0.450324976, \quad 0.493884786.$$

Solution to Q16.15. We are given ν so we first compute

$$\mu = \nu - 0.5\sigma^2 = 0.095.$$

To get T we divide by 365 to get 0.02740.

We compute the inverse cumulative normals of the VAR levels to get

$$-3.0902, \quad -2.3263, \quad -1.6449.$$

We then just plug into the formula for VAR

$$4.7388, \quad 3.5266, \quad 2.4322.$$

Solution to Q16.16. We are given ν so we first compute

$$\mu = \nu - 0.5\sigma^2 = 0.005.$$

To get T we divide by 365 to get 0.00274.

We compute the inverse cumulative normals of the VAR levels to get

$$-3.0902, \quad -2.3263, \quad -1.6449.$$

We then just plug into the formula for VAR

$$4.7353, \quad 3.5858, \quad 2.5485.$$

Solution to Q16.17. We are given ν so we first compute

$$\mu = \nu - 0.5\sigma^2 = 0.005.$$

To get T we divide by 365 to get 0.082191781.

We compute the inverse cumulative normals of the VAR levels to get

$$-3.0902, \quad -2.3263, \quad -1.6449.$$

We then just plug into the formula for VAR

$$51.0567, \quad 41.6848, \quad 31.8191.$$

Solution to Q16.18. We add 25 to the initial and terminal values of the stock price and then proceed as for a log-normal random variable.

We have

$$T = 0.027397, \quad \mu = 0.08, \quad S_0 + D = 125.$$

The VARs are

$$11.9075, \quad 9.0112, \quad 6.3647.$$

Solution to Q16.19. We add 25 to the initial and terminal values of the stock price and the level targeted, and then proceed as for a log-normal random variable.

We have

$$\mu = 0.01875, \quad S_0 + D = 110.$$

The probabilities are

$$0.392409117, \quad 0.444314461, \quad 0.471682818.$$

References

[1] M.E. Blume, On the assessment of risk, *Journal of Finance*, 1971, **26** (1), 1–10.

[2] N-F. Chen, R. Roll and S. Ross, *Journal of Business*, 1986, **59** (3), 383–403.

[3] J.H. Cochrane, *Asset Pricing,* Princeton University Press, 2001.

[4] W.F.M. De Bondt, R.H. Thaler, Does the stock market overreact? *Journal of Finance*, 1985, **40**, 793–805.

[5] E. Elton, M.J. Gruber, Estimating the dependence of share prices – implications for portfolio selection, *Journal of Finance*, 1973, **VIII** (5), 1203–1322.

[6] E.J. Elton, M.J. Gruber, S.J. Brown, W.N. Goetzmann, *Modern Portfolio Theory and Investment Analysis,* Wiley, 6th ed., 2003.

[7] R. Harris, The Bubble Act: its passage and its effects on business organization, *Journal of Economic History*, 1994, **54** (3), 610–627 Published by: Cambridge University Press on behalf of the Economic History Association, Stable URL: `http://www.jstor.org/stable/2123870`.

[8] M. Joshi, *The Concepts and Practice of Mathematical Finance,* Cambridge University Press, 2nd ed., 2008.

[9] J.L. Kelly, A new interpretation of information rate, 1956, *Bell System Technical Journal,* **35**, 917–926.

[10] G. Mackay, *Extraordinary Popular Delusions and the Madness of Crowds*, Office of the National Illustrated Library, London, 1841.

[11] H.M. Markowitz, *Portfolio Selection,* Blackwell Publishing, 2nd ed., 1990.

[12] R.C. Merton, An analytical derivation of the efficient portfolio frontier, *Journal of Financial and Quantitative Analysis,* 1972, **7**, 1851–1872,

[13] M.H. Miller, M. Scholes, Rates of return in relation to risk: a re-examination of some recent findings. In *Studies in the Theory of Capital Markets,* Praeger, 1972.

[14] J. von Neumann, O. Morgenstern, *Theory of Games and Economic Behavior,* Princeton University Press, 1944; 2nd ed. 1947; 3rd ed. 1953.

[15] G. Pennachi, *Theory of Asset Pricing,* Pearson Education, 2008.

[16] M. Rabin, R. Thaler, Anomalies: risk aversion, *Journal of Economic Perspectives,* 2001, **15**, (1), 219–232.

[17] R. Roll, A critique of the asset pricing theory's tests Part I: On past and potential testability of the theory, *Journal of Financial Economics*, 1997, **4**, (2), 129–176.

[18] S. Ross, *Neoclassical Finance,* Princeton University Press, 2004.

[19] P.A. Samuelson, The "fallacy" of maximizing the geometric mean in long sequences of investing or gambling, *Proceedings of the National Academy of Sciences,* 1971, **66** 2493–2496.

[20] R. Shiller, *Irrational Exuberance,* Crown Business, 2nd ed., 2006.

[21] J. Shanken, Multivariate proxies and asset pricing relations: Living with the Roll critique, *Journal of Financial Economics*, 1987, **18**, (1), 91–110.

[22] W.F. Sharpe, G.M. Cooper, Risk-return classes of New York Stock Exchange common stocks 1931–1967, *Financial Analysts Journal,* 1972, **28** (2), 46+48–54+81.

[23] A. Shleifer, *Inefficient Markets: an Introduction to Behavioral Finance,* Oxford University Press, 2000.

[24] A. Smithers, *Wall Street Revalued: Imperfect Markets and Inept Central Bankers,* Wiley, 2009.

[25] N. Taleb, *Fooled by Randomness*, Random House, 2008.

[26] N. Taleb, *The Black Swan*, Random House, 2007.

[27] E.A. Thompson, J. Treussard, The Tulipmania: fact or artifact, Levine's Working Paper Archive 618897000000000830, David K. Levine, 2003.

[28] O.A. Vasicek, A note on using cross-sectional information in Bayesian estimation of security betas, *Journal of Finance,* 1973, **28** (5), 1233–1239.

Index